Lecture Notes
in Control and Infor

Editors: M. Thoma · M. Morari

Michael I. Gil'

Explicit Stability Conditions for Continuous Systems

A Functional Analytic Approach

Author

Prof. Dr. Michael I. Gil'
Ben Gurion University of Negev
Department of Mathematics
P.O. Box 653
84 105 Beer Sheva
Israel

ISSN 0170-8643

ISBN 3-540-23984-7 **Springer Berlin Heidelberg New York**

Library of Congress Control Number: 2005920065

Springer is a part of Springer Science+Business Media

springeronline.com

Printed in The Netherlands

Typesetting: Data conversion by the authors.
Final processing by PTP-Berlin Protago-TEX-Production GmbH, Germany
Cover-Design: design & production GmbH, Heidelberg
Printed on acid-free paper 89/3141Yu - 5 4 3 2 1 0

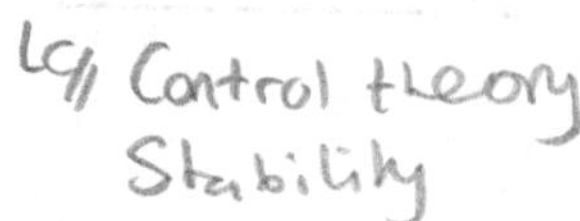

Preface

This book deals with nonautonomous linear and nonlinear continuous finite dimensional systems. Explicit conditions for the asymptotic, absolute, input-to-state and orbital stabilities are discussed.

The problem of stability analysis of various systems continues to attract the attention of many specialists despite its long history. It is still one of the most burning problems of control theory, because of the absence of its complete solution. The problem of the synthesis of a stable system is closely connected with the problem of stability analysis. Any progress in the problem of analysis implies success in the problem of synthesis of stable systems.

The basic method for the stability analysis of nonlinear systems is the Lyapunov functions one. By this method many very strong results are obtained, but finding Lyapunov's functions is often connected with serious mathematical difficulties, especially in regard to nonstationary systems. The stability conditions presented in this book are mainly formulated in terms of the eigenvalues of auxiliary matrices. This fact allows us to apply the well-known stability criteria for polynomials and matrices (for example the Hurwitz criterion) to the stability analysis of time-varying (linear and nonlinear) systems.

The main methodology presented in this publication is based on a combined use of recent norm estimates for matrix-valued functions with the following well-known methods and results:

a) the method of characteristic exponents (the first Lyapunov method);
b) the multiplicative representations of solutions;
c) the freezing method;
d) the positivity of the Green (impulse) functions.

Here we do not consider the Lyapunov functions method because several excellent books cover this topic.

A significant part of this book is devoted to a solution to the problem connected with the Aizerman conjecture. Recall that in 1949 M. A. Aizerman conjectured that a single input-single output system is absolutely stable in the Hurwitz angle. This hypothesis caused great interest among the specialists. With the help of counter examples it was shown that the conjecture is not true, in general. The problem of finding a class of systems that satisfy Aizerman's hypothesis arose. One of the most powerful results in this direction was obtained in 1966 by N. Truchan who showed that Aizerman's hypothesis is satisfied by systems having linear parts in the form of single loop circuits with up to five stable aperiodic links connected in tandem. In 1981 the author showed that any system satisfies the Aizerman hypothesis if its Green function is non-negative. That result includes the Truchan one. Similar results are derived for multivariable systems. They are also presented here.

The aim of this book is to provide new tools for specialists in control system theory and stability theory of ordinary differential equations. We suggest a systematic exposition of the approach to stability analysis which is based on estimates for matrix-valued functions. That approach allows us to investigate various classes of systems from the unified viewpoint.

The book is intended not only for specialists in stability theory, but for anyone interested in various applications who has had at least a first year graduate level course in analysis.

I was very fortunate to have fruitful discussions with Professors M.A. Aizerman, V.M. Alekseev, M.A. Krasnosel'skii, A. Pokrovskii and A.A. Voronov, to whom I am very grateful for their interest in my investigations.

Acknowledgment. This work was supported by the Kamea Fund of the Israel.

Michael I. Gil'

Table of Contents

Introduction

The topic of this book is the stability analysis of continuous finite dimensional systems. The book consists of 13 chapters and two appendices.

In Chapter 1 some well-known and preliminary results are collected. They are systematically used in the next chapters. In particular, in Sect. 1.5, we review the estimates for the norms of matrix-valued functions. They are the main tool of our investigations.

Denote by $\mathbf{C}^n$ and $\mathbf{R}^n$ the complex and real Euclidean spaces, respectively; I denotes the unit matrix, $\|.\|$ is the Euclidean norm. Let A be an $n \times n$-matrix. *The following quantity plays an essential role in the book:*

$$g(A) = (N^2(A) - \sum_{k=1}^{n} |\lambda_k(A)|^2)^{1/2},$$

where $N(A)$ is the Frobenius (Hilbert - Schmidt) norm of A, and $\lambda_k(A)$ $(k = 1, ..., n)$ are the eigenvalues of A taken with their multiplicities. The relations

$$g^2(A) \leq N^2(A) - |Trace A^2| \text{ and } g^2(A) \leq N^2(A - A^*)/2$$

are true. Here A^* is the adjoint matrix. Put $\alpha(A) = \max_{k=1,\dots,n} Re\ \lambda_k(A)$. The estimate

$$\|exp(At)\| \leq e^{\alpha(A)t} \sum_{k=0}^{n-1} \frac{g^k(A)t^k}{(k!)^{3/2}} \quad (t \geq 0) \tag{1}$$

is often applied in the book. If A is a normal matrix, then $g(A) = 0$, and $\|exp(At)\| = e^{\alpha(A)t}$.

We also use is the following estimates for the resolvent:

$$\|(A - \lambda I)^{-1}\| \leq \sum_{k=0}^{n-1} \frac{g^k(A)}{\sqrt{k!}\rho^{k+1}(A, \lambda)} \text{ for all regular } \lambda,$$

where $\rho(A, \lambda)$ is the distance between the spectrum of A and a regular point λ, and

$$\|(I\lambda - A)^{-1}\| \leq \frac{N^{n-1}(A - I\lambda)}{|det\ (\lambda I - A)|\ (n-1)^{(n-1)/2}},$$

where *det* denotes the determinant.

Chapters 2–10 are devoted to asymptotic, exponential and absolute stabilities of linear and nonlinear systems.

In Chap. 2 we consider perturbations of the linear system

$$\dot{x}(t) = A(t)x(t) \;\; (\dot{x}(t) \equiv dx(t)/dt, \; t \geq 0) \tag{2}$$

with a variable matrix $A(t)$. In addition, the multiplicative representation for solutions is introduced. That representation is used for constructing of majorants and minorants of solutions. Moreover, we establish stability conditions for systems, which are close to triangular ones.

In Chap. 3 we investigate linear system (2) in the case when $A(t)$ is a slowly varying matrix. To illustrate a typical result assume that the matrix $A(t)$ satisfies the condition

$$\|A(t) - A(s)\| \leq q_0|t - s| \;\; (t, s \geq 0), \tag{3}$$

where q_0 is a positive constant. In addition,

$$v := \sup_{t \geq 0} g(A(t)) < \infty. \tag{4}$$

Denote by $z(q_0, v)$ the extreme right-hand (positive) root of the algebraic equation $z^{n+1} = q_0 P(z)$, where

$$P(z) = \sum_{k=0}^{n-1} \frac{(k+1)v^k}{\sqrt{k!}} z^{n-k-1}.$$

By the freezing method and estimate (1), the following result is proved: let conditions (3) and (4) hold. In addition, let the matrix $A(t) + z(q_0, v)\, I$ be a Hurwitz one for all $t \geq 0$. Then system (2) is stable.

We also give simple estimates for $z(q_0, v)$.

In Chap. 4 we continue to investigate linear multivariable systems. Stability conditions are derived for systems with piecewise constant matrices. Moreover, by the multiplicative representation, the Lozinskii and Wazewski inequalities are proved. In addition the second order vector equation

$$\ddot{x} + A(t)\dot{x} + B(t)x = 0$$

is considered. Here $A(t), B(t)$ are variable matrices.

Chapter 5 is devoted to the system

$$\dot{x} = Ax + F(x, t) \; (t \geq 0), \tag{5}$$

where A is a constant $n \times n$-matrix and F maps $\mathbf{C}^n \times [0, \infty)$ into $\mathbf{C}^n$ with the property

$$\|F(h, t)\| \leq \nu \|h\| \text{ for all } h \in \Omega(r) \text{ and } t \geq 0. \tag{6}$$

Here $\nu = const \geq 0$, and $\Omega(r) = \{h \in \mathbf{C}^n : \|h\| \leq r\}$ $(r \leq \infty)$. By estimate (1) and the first Lyapunov method, conditions for the exponential stability of system (5) are established.

Chapter 6 is devoted to the Aizerman conjecture. Let us consider in $\mathbf{R}^n$ the equation

$$\dot{y} = Ay + b\, f(s,t) \; (s = cy, \; t \geq 0), \tag{7}$$

where A is a constant Hurwitz $n \times n$-matrix, b is a column, c is a row, f maps $\mathbf{R}^1 \times [0.\infty)$ into $\mathbf{R}^1$ with the property

$$0 \leq f(s,t)/s \leq q \; (s \in \mathbf{R}^1, \; s \neq 0, \; t \geq 0). \tag{8}$$

In 1949 M. A. Aizerman formulated the following conjecture: under the condition $f(s,t) \equiv f(s)$, for the absolute stability of the zero solution of system (7) in the class of nonlinearities (8) it is necessary and sufficient that the linear equation

$$\dot{y} = Ay + q_1 bcy$$

be asymptotically stable for any $q_1 \in [0, q]$. This conjecture is not, in general, true. Therefore, the following problem arose: to find the class of systems that satisfy Aizerman's hypothesis. To formulate the relevant result let us introduce the transfer function of the linear part of system (7):

$$W(\lambda) = c(\lambda I - A)^{-1} b = \frac{L(\lambda)}{P(\lambda)}.$$

Here $P(\lambda)$ and $L(\lambda)$ are polynomials. Besides, let

$$K(t) := \frac{1}{2\pi} \int_{-\infty}^{\infty} e^{iyt} W(iy) dy.$$

That is, $K(t)$ is the corresponding Green (impulse) function. It is proved that under the condition

$$K(t) \geq 0 \; (t \geq 0),$$

the zero solution of system (7) is absolutely stable in the class of nonlinearities (8) if and only if the polynomial $P(\lambda) - qL(\lambda)$ is Hurwitzian. Clearly, that result singles out one of the classes of linear parts of systems that satisfy the Aizerman conjecture. It is also proved that the polynomial $P(\lambda) - qL(\lambda)$ is Hurwitzian, provided $P(0) > qL(0)$ and the Green function is positive. Moreover, Chapter 6 also contains the generalized Aizerman problem for multivariable systems.

Chapter 7 deals with nonlinear systems of the type

$$\dot{x} = A(t)x + F(x,t) \;\; (t \geq 0), \tag{9}$$

where $A(t)$ is a variable $n \times n$-matrix and F maps $\mathbf{C}^n \times [0, \infty)$ into $\mathbf{C}^n$ with the property (6).

Chapter 8 deals with nonlinear systems of the type

$$\dot{x}(t) = B(x(t), t)x(t) \quad (t \geq 0), \tag{10}$$

where

$$B(h, t) = (b_{jk}(h, t))_{j,k=1}^{n}$$

is an $n \times n$-matrix for every $h \in \mathbf{C}^n$ and $t \geq 0$. In particular, the freezing method is developed to nonlinear systems. In addition, nonlinear dissipative systems, nonlinear triangular systems and their perturbations are investigated. Moreover, the very interesting Levin stability criterion for nonlinear scalar equations with real variable characteristic roots is presented. In Chapter 8 we also consider systems of the type

$$\ddot{x} + F(x, \dot{x}, t)\dot{x} + G(x, \dot{x}, t)x = 0,$$

where $F(h, w, t)$ and $G(h, w, t)$ are matrices continuously dependent on $h, w \in \mathbf{R}^n$ and $t \geq 0$.

In Chapter 9 we consider the Lur'e type systems

$$\ddot{x} + A\dot{x} + Bx = \Phi(x, t),$$

where A and B are constant matrices, and Φ is a continuous vector-valued function. Stability conditions and bounds for the region of attraction are derived.

One contour nonlinear nonautonomous systems with positive Green functions are investigated in Chap. 10.

Chapter 11 is devoted to input-to-state stability of nonlinear systems. Here we consider the input-to-state version of Aizerman's conjecture.

Orbital stability and forced oscillations are discussed in Chap. 12. In particular, conditions that provide the existence of forced oscillations are established.

Conditions for the existence of positive and nontrivial steady states are investigated in Chap. 13.

In Appendix A, some well known bounds for eigenvalues of matrices are collected. In Appendix B, positivity conditions for the Green functions of ordinary differential equations are presented.

1. Preliminaries

1.1 Vector and Matrix Norms

In the sequel, $\mathbf{C}^n$ is an n-dimensional complex Euclidean space and

$$\|x\| = \sqrt{\sum_{k=1}^{n} |x_k|^2} \ .$$

is the Euclidean norm of $x = (x_k) \in \mathbf{C}^n$. Recall the following properties of the norm:

$\|x\| = 0$ iff $x = 0$,
$\|\alpha x\| = |\alpha| \|x\|$,
$\|x + y\| \le \|x\| + \|y\|$ for all $x, y \in \mathbf{C}^n$, $\alpha \in \mathbf{C}$,

$$\|x - y\| \ge \|x\| - \|y\| \text{ and } \|x\| = \| - x\|.$$

We will use the following matrix norms: the operator norm and the Frobenius (Hilbert-Schmidt) norm. The operator norm of a matrix (a linear operator in $\mathbf{C}^n$) A is

$$\|A\| = \sup_{x \in C^n} \frac{\|Ax\|}{\|x\|}.$$

The relations

$$\|A\| > 0 \ (A \ne 0); \ \|\lambda A\| = |\lambda| \|A\| \ (\lambda \in \mathbf{C}),$$

$$\|AB\| \le \|A\| \|B\| \text{ and } \|A + B\| \le \|A\| + \|B\|$$

are valid for all matrices A and B. The Frobenius norm of A is

$$N(A) = \sqrt{\sum_{j,k=1}^{n} |a_{jk}|^2}.$$

Here a_{jk} are the entries of matrix A in some orthogonal normal basis. Moreover,

$$N^2(A) = Trace \ (AA^*),$$

where $A^* = (\overline{a}_{kj})_{j,k=1}^n$ is the adjoint matrix. Thus, the Frobenius norm does not depend on the choice of an orthogonal normal basis. The relations

$$N(A) > 0 \ (A \neq 0); \ N(\lambda A) = |\lambda| N(A) \ (\lambda \in \mathbf{C}),$$

$$N(AB) \leq N(A)N(B) \text{ and } N(A+B) \leq N(A) + N(B)$$

are true for all matrices A and B. Furthermore, $\mathbf{R}^n$ denotes the real Euclidean space, I is the unit matrrix and $(.,.)$ is the scalar product in $\mathbf{R}^n$ or $\mathbf{C}^n$. So

$$\|x\| = \sqrt{(x,x)}.$$

1.2 Definitions of Stability

Put $R_+ = [0, \infty)$ and consider in $\mathbf{C}^n$ the differential equation

$$\dot{x}(t) = f(t, x(t)) \ \ (t > 0; \ \dot{x}(t) \equiv dx/dt), \tag{2.1}$$

where $f : R_+ \times \mathbf{C}^n \to \mathbf{C}^n$ is a continuous vector-valued function. *A solution* of (2.1) is a differentiable function $x : R_+ \to \mathbf{C}^n$ satisfying that equation for all $t > 0$.

It is further assumed that the function f is of such a nature that equation (2.1) has a unique solution over R_+ corresponding to each initial condition $x(0) = x_0$. For example, if f satisfies the Lipshitz condition (see for instance (Vidyasagar, 1993, Theorem 2.4.25)).

A point $x_1 \in \mathbf{C}^n$ is said to be *an equilibrium point* of the system (2.1) if

$$f(t, x_1) \equiv 0 \text{ for all } t \geq 0.$$

In other words, if the system starts at an equilibrium point, it stays there. The converse is also true. Throughout this book we assume that 0 is an equilibrium point of the system (2.1). This assumption does not result in any loss of generality, because if x_1 is an equilibrium point of (2.1), then 0 is an equilibrium point of the system

$$\dot{z}(t) = f_1(t, z(t)),$$

where $f_1(t, z(t)) = f(t, z(t) + x_1)$. Everywhere below in this section, $x(t)$ is a solution of (2.1).

Definition 1.2.1 *The equilibrium point* 0 *is said to be stable (in the sense of Lyapunov) if, for every* $t_0 \geq 0$ *and* $\epsilon > 0$, *there exists* $\delta(t_0) > 0$, *such that the condition* $\|x(t_0)\| \leq \delta(t_0)$ *implies*

$$\|x(t)\| \leq \epsilon \ (t \geq t_0). \tag{2.2}$$

It is uniformly stable if, for each $\epsilon > 0$, *there exists* $\delta > 0$ *independent of* t_0, *such that the condition* $\|x(t_0)\| \leq \delta$ *implies inequality (2.2).*

The equilibrium is *unstable*, if it is not stable.

Because all norms on $\mathbf{C}^n$ are topologically equivalent, it follows that the stability status of an equilibrium does not depend on the particular norm.

Definition 1.2.2 *The equilibrium 0 is asymptotically stable if it is stable and for each $t_0 \in R_+$, there is an $\eta(t_0) > 0$ such that*

$$\|x(t_0)\| \leq \eta(t_0) \textit{ implies } x(t) \to 0 \textit{ as } t \to \infty.$$

The equilibrium 0 is uniformly asymptotically stable if it is stable and there is an $\eta > 0$ independent of t_0, such that

$$\|x(t_0)\| \leq \eta \textit{ implies } x(t) \to 0 \textit{ as } t \to \infty.$$

Also, the set $\Omega(\eta)$, defined by

$$\Omega(\eta) = \{x \in \mathbf{C}^n : \|x\| \leq \eta\}$$

is called the region of attraction (the stability domain) for the equilibrium point 0.

The equilibrium point 0 is said to be globally (uniformly) asymptotically stable if it is (uniformly) asymptotically stable and the region of attraction $\Omega(\eta) = \mathbf{C}^n$.

As shown by Vinograd (1957) (see also (Vidyasagar, 1993, p. 141)), attractivity and stability are really independent properties, i.e., an equilibrium can be attractive without being stable.

Definition 1.2.3 *The equilibrium point 0 is exponentially stable if for any $t_0 \geq 0$ there exist constants $r, a, b > 0$ such that*

$$\|x(t)\| \leq a\|x(t_0)\|e^{-b(t-t_0)} \; (t \geq t_0), \; \textit{if } \|x(t_0)\| \leq r. \tag{2.3}$$

It is uniformly exponentially stable if constants $r, a, b > 0$ in (2.3) are independent of t_0. The equilibrium point 0 is globally exponentially stable if inequality (2.3) holds with $r = \infty$.

Consider the linear equation

$$\dot{u}(t) = A(t)u(t) \; (t \geq 0) \tag{2.4}$$

with a variable $n \times n$-matrix $A(t)$. Since the zero is the unique equilibrium point of a linear equation, *we will say that (2.4) is stable (uniformly stable, asymptotically stable, exponentially stable)* if the zero solution of (2.4) is stable (uniformly stable, asymptotically stable, exponentially stable).

For linear equations, the notions of the global asymptotic (exponential) stability and asymptotic (exponential) stability coincide.

In addition, *we will say that (2.4) is uniformly asymptotically (exponentially) stable* if the zero solution of (2.4) is uniformly asymptotically (exponentially) stable.

1.3 Eigenvalues of Matrices

Recall that for an $n \times n$-matrix (a linear operator in $\mathbf{C}^n$) $A = (a_{jk})_{j,k=1}^n$, $A^* = (\overline{a}_{kj})_{j,k=1}^n$, is the adjoint matrix. In other words

$$(Ax, y) = (x, A^*y) \ (x, y \in \mathbf{C}^n).$$

A is a Hermitian matrix if $A^* = A$. It is positive definite if it is Hermitian and $(Ah, h) \geq 0$ for any nonzero $h \in \mathbf{C}^n$;

A is a normal matrix if $AA^* = A^*A$. It is nilpotent if $A^n = 0$.

Let A be an arbitrary matrix. Then the matrices

$$A_I = (A - A^*)/2i \text{ and } A_R = (A + A^*)/2$$

are the imaginary Hermitian component and real Hermitian one of A, respectively. By A^{-1} the matrix inverse to A is denoted: $AA^{-1} = A^{-1}A = I$.

Let A be an arbitrary matrix. Then if for some $\lambda \in \mathbf{C}$, the equation

$$Ah = \lambda h$$

has a nontrivial solution, λ is an eigenvalue of A and h is its eigenvector. An eigenvalue $\lambda(A)$ has the (algebraic) multiplicity r if

$$dim(\cup_{k=1}^n ker(A - \lambda I)^k) = r.$$

Let $\lambda_k(A)$ $(k = 1, \ldots, n)$ be eigenvalues of A, including with their multiplicities. Then the set

$$\sigma(A) = \{\lambda_k(A)\}_{k=1}^n$$

is the spectrum of A.

All eigenvalues of a Hermitian matrix A are real. If, in addition, A is positive (negative) definite, then all its eigenvalues are positive (negative). Furthermore,

$$r_s(A) = \max_{k=1,\ldots,n} |\lambda_k(A)|$$

is the spectral radius of A. Denote

$$\alpha(A) = \max_{k=1,\ldots,n} Re\lambda_k(A), \ \beta(A) = \min_{k=1,\ldots,n} Re\lambda_k(A).$$

A matrix A *is said to be a Hurwitz one* if all its eigenvalues lie in the open left half-plane, i.e., $\alpha(A) < 0$.

A complex number λ is *a regular point of A* if it does not belong to the spectrum of A, i.e., if $\lambda \neq \lambda_k(A)$ for any $k = 1, \ldots, n$.

Recall some properties of the trace $Trace\ A = Tr\ A$ of A :

$$Tr\ A = \sum_{k=1}^n a_{kk} = \sum_{k=1}^n \lambda_k(A).$$

In addition $Tr\ (A+B) = Tr\ A + Tr\ B$ for all matrices A and B. By $det(A)$ the determinant of A is denoted:

$$det(A) = \prod_{k=1}^{n} \lambda_k(A).$$

A polynomial

$$p(\lambda) = det(\lambda I - A) = \prod_{k=1}^{n} (\lambda - \lambda_k(A))$$

is said to be the characteristic polynomial of A. All the eigenvalues of A are the roots of its characteristic polynomial. The algebraic multiplicity of an eigenvalue of A coincides with the multiplicity of the corresponding root of the characteristic polynomial. A polynomial $P(\lambda)$ is said to be *a Hurwitz one* if all its roots lie in the open left half-plane. Thus, the characteristic polynomial of a Hurwitz matrix is a Hurwitz polynomial.

1.4 Matrix-Valued Functions

Let A be a matrix and let $f(\lambda)$ be a scalar-valued function which is analytical on a neighborhood D of $\sigma(A)$. We define the function $f(A)$ of A by the generalized integral formula of Cauchy

$$f(A) = -\frac{1}{2\pi i} \int_{\Gamma} f(\lambda) R_\lambda(A) d\lambda,$$

where $\Gamma \subset D$ is a closed smooth contour surrounding $\sigma(A)$, and

$$R_\lambda(A) = (A - \lambda I)^{-1}$$

is the resolvent of A. If an analytic function $f(\lambda)$ is represented in the domain

$$\{z \in \mathbf{C} : |z| \le r_s(A)\}$$

by the Taylor series

$$f(\lambda) = \sum_{k=0}^{\infty} c_k \lambda^k,$$

then

$$f(A) = \sum_{k=0}^{\infty} c_k A^k.$$

In particular, for any matrix A,

$$e^A = \sum_{k=0}^{\infty} \frac{A^k}{k!}.$$

Example 1.4.1 *Let A be a diagonal matrix:*

$$A = \begin{pmatrix} a_1 & 0 & \dots & 0 \\ 0 & a_2 & \dots & 0 \\ . & \dots & . & . \\ 0 & \dots & 0 & a_n \end{pmatrix}.$$

Then

$$f(A) = \begin{pmatrix} f(a_1) & 0 & \dots & 0 \\ 0 & f(a_2) & \dots & 0 \\ . & \dots & . & . \\ 0 & \dots & 0 & f(a_n) \end{pmatrix}.$$

Example 1.4.2 *If a matrix J is an $n \times n$-Jordan block:*

$$J = \begin{pmatrix} \lambda_0 & 1 & 0 & \dots & 0 \\ 0 & \lambda_0 & 1 & \dots & 0 \\ . & . & . & \dots & . \\ . & . & . & \dots & . \\ . & . & . & \dots & . \\ 0 & 0 & \dots & \lambda_0 & 1 \\ 0 & 0 & \dots & 0 & \lambda_0 \end{pmatrix},$$

then

$$f(J) = \begin{pmatrix} f(\lambda_0) & \frac{f'(\lambda_0)}{1!} & \dots & \frac{f^{(n-1)}(\lambda_0)}{(n-1)!} \\ 0 & f(\lambda_0) & \dots & \\ . & . & \dots & . \\ . & . & \dots & . \\ . & . & \dots & . \\ 0 & \dots & f(\lambda_0) & \frac{f'(\lambda_0)}{1!} \\ 0 & \dots & 0 & f(\lambda_0) \end{pmatrix}.$$

1.5 Norms of Matrix Functions

Let $A = (a_{jk})$ be an $n \times n$-matrix. *The following quantity plays a key role in the sequel:*

$$g(A) = (N^2(A) - \sum_{k=1}^{n} |\lambda_k(A)|^2)^{1/2}. \tag{5.1}$$

Recall that I is the unit matrix, $N(A)$ is the Frobenius (Hilbert-Schmidt) norm of A, and $\lambda_k(A)$ $(k = 1, ..., n)$ are the eigenvalues taken with their multiplicities. Since

$$\sum_{k=1}^{n} |\lambda_k(A)|^2 \geq |Trace\ A^2|,$$

we get

$$g^2(A) \leq N^2(A) - |Trace\ A^2|. \tag{5.2}$$

In (Gil, 2003, Section 2.1), the following relations are proved:

$$g^2(A) \leq \frac{1}{2} N^2(A^* - A) \tag{5.3}$$

and

$$g(e^{i\tau} A + zI) = g(A) \tag{5.4}$$

for all $\tau \in \mathbf{R}$ and $z \in \mathbf{C}$. If A is a normal matrix: $AA^* = A^*A$, then $g(A) = 0$. To formulate the result, for a natural $n > 1$ introduce the numbers

$$\gamma_{n,k} = \sqrt{\frac{C_{n-1}^k}{(n-1)^k}} \quad (k = 1, ..., n-1) \text{ and } \gamma_{n,0} = 1.$$

Here

$$C_{n-1}^k = \frac{(n-1)!}{(n-k-1)!k!}$$

are binomial coefficients. Evidently, for all $n > 2$,

$$\gamma_{n,k}^2 = \frac{(n-1)(n-2)\dots(n-k)}{(n-1)^k k!} \leq \frac{1}{k!} \quad (k = 1, 2, ..., n-1). \tag{5.5}$$

Let $B(\mathbf{C}^n)$ be the set of all linear operators (matrices) in $\mathbf{C}^n$.

Theorem 1.5.1 *Let $A \in B(\mathbf{C}^n)$ and let f be a function regular on a neighborhood of the closed convex hull $co(A)$ of the eigenvalues of A. Then*

$$\|f(A)\| \leq \sum_{k=0}^{n-1} \sup_{\lambda \in co(A)} |f^{(k)}(\lambda)| g^k(A) \frac{\gamma_{n,k}}{k!}.$$

The proof of this theorem can be found in (Gil, 2003, Theorem 2.7.1). This theorem is exact: if A is a normal matrix and

$$\sup_{\lambda \in co(A)} |f(\lambda)| = \sup_{\lambda \in \sigma(A)} |f(\lambda)|,$$

then we have the equality $\|f(A)\| = \sup_{\lambda \in \sigma(A)} |f(\lambda)|$.

The previous theorem and (5.5) give us the following

Corollary 1.5.2 *Let $A \in B(\mathbf{C}^n)$ and let f be a function regular on a neighborhood of the closed convex hull $co(A)$ of the eigenvalues of A. Then*

$$\|f(A)\| \leq \sum_{k=0}^{n-1} \sup_{\lambda \in co(A)} |f^{(k)}(\lambda)| \frac{g^k(A)}{(k!)^{3/2}}.$$

Theorem 1.5.1 and Corollary 1.5.2 give us the following

Corollary 1.5.3 *For a linear operator A in $\mathbf{C}^n$,*

$$\|exp(At)\| \leq e^{\alpha(A)t} \sum_{k=0}^{n-1} g^k(A) t^k \frac{\gamma_{n,k}}{k!} \leq e^{\alpha(A)t} \sum_{k=0}^{n-1} \frac{g^k(A)t^k}{(k!)^{3/2}} \quad (t \geq 0)$$

where $\alpha(A) = \max_{k=1,\ldots,n} Re\,\lambda_k(A)$. In addition,

$$\|A^m\| \leq \sum_{k=0}^{n-1} \frac{\gamma_{n,k} m! g^k(A) r_s^{m-k}(A)}{(m-k)!k!} \leq \sum_{k=0}^{n-1} \frac{m! g^k(A) r_s^{m-k}(A)}{(m-k)!(k!)^{3/2}} \quad (m = 1, 2, \ldots)$$

where $r_s(A)$ is the spectral radius. Recall that $1/(m-k)! = 0$ if $m < k$.

Furtermore we will use the following result.

Corollary 1.5.4 *Let A be an $n \times n$-matrix. Then for any $h \in \mathbf{C}^n$, the inequality*

$$\|exp(At)h\| \geq e^{[\beta(A)t]} \, [\sum_{k=0}^{n-1} g^k(A) t^k (k!)^{-1} \gamma_{n,k}]^{-1} \, \|h\| \ (t \geq 0)$$

is valid. Therefore, in the accordance with (5.5),

$$\|exp(At)h\| \geq e^{[\beta(A)t]} [\sum_{k=0}^{n-1} g^k(A)(k!)^{-3/2} \, t^k]^{-1} \, \|h\| \ (t \geq 0).$$

Indeed, since

$$\max Re\sigma(-A) = -\min Re\sigma(A) = -\beta(A),$$

and $g(-A) = g(A)$, Corollary 1.5.3 yields

$$\|exp(-At)v\| \leq \|v\| e^{-\beta(A)t} \sum_{k=0}^{n-1} g^k(A) t^k \frac{\gamma_{n,k}}{k!} \ (t \geq 0)$$

for any $v \in \mathbf{C}^n$. Taking into account that the operator $exp(-At)$ is the inverse one to $exp(At)$ and putting $exp(-At)v = h$, we arrive at the assertion.

We will need also the following results.

Theorem 1.5.5 *Let A be a linear operator in $\mathbf{C}^n$. Then its resolvent $R_\lambda(A) = (A - \lambda I)^{-1}$ satisfies the inequality*

$$\|R_\lambda(A)\| \leq \sum_{k=0}^{n-1} \frac{g^k(A)\gamma_{n,k}}{\rho^{k+1}(A, \lambda)} \text{ for any regular point } \lambda \text{ of } A,$$

where $\rho(A, \lambda) = \min_{k=1,\ldots,n} |\lambda - \lambda_k(A)|$.

The proof of this theorem can be found in (Gil, 2003, Section 2.1). Theorem 1.5.5 is exact: if A is a normal matrix, then $g(A) = 0$ and

$$\|R_\lambda(A)\| = \frac{1}{\rho(A, \lambda)} \text{ for all regular points } \lambda \text{ of } A.$$

Let A be an invertible $n \times n$-matrix. Then by Theorem 1.5.5,

$$\|A^{-1}\| \leq \sum_{k=0}^{n-1} g^k(A) \frac{\gamma_{n,k}}{\rho_0^{k+1}(A)},$$

where $\rho_0(A) = \rho(A, 0)$ is the smallest modulus of the eigenvalues of A:

$$\rho_0(A) = \inf_{k=1,\ldots,n} |\lambda_k(A)|.$$

Moreover, Theorem 1.5.5 and inequalities (5.5) imply

Corollary 1.5.6 *Let A be a linear operator in $\mathbf{C}^n$. Then*

$$\|R_\lambda(A)\| \leq \sum_{k=0}^{n-1} \frac{g^k(A)}{\sqrt{k!}\rho^{k+1}(A, \lambda)} \text{ for any regular point } \lambda \text{ of } A.$$

The following result is proved in (Gil', 2003, Section 2.11).

Lemma 1.5.7 *Let A be $n \times n$-matrix $(n > 1)$. Then*

$$\|(I\lambda - A)^{-1}\| \leq \frac{(N^2(A) - 2Re\,(\overline{\lambda}\, Trace\,(A)) + n|\lambda|^2)^{(n-1)/2}}{|det\,(\lambda I - A)|\ (n-1)^{(n-1)/2}} =$$

$$\frac{N^{n-1}(A - I\lambda)}{|det\,(\lambda I - A)|\ (n-1)^{(n-1)/2}}$$

for any regular λ of A.

1.6 Examples

In this section we present some examples of calculations of $g(A)$.

Example 1.6.1 *Consider the matrix*

$$A = \begin{pmatrix} a_{11} & a_{12} \\ a_{21} & a_{22} \end{pmatrix}$$

where a_{jk} $(j, k = 1, 2)$ are real numbers. Then due to (5.3)

$$g(A) \leq |a_{12} - a_{21}|.$$

Now let us consider the case of *nonreal eigenvalues*: $\lambda_2(A) = \overline{\lambda}_1(A)$. It can be written

$$det(A) = \lambda_1(A)\overline{\lambda}_1(A) = |\lambda_1(A)|^2$$

and

$$|\lambda_1(A)|^2 + |\lambda_2(A)|^2 = 2|\lambda_1(A)|^2 = 2det(A) = 2[a_{11}a_{22} - a_{21}a_{12}].$$

Thus,

$$g^2(A) = N^2(A) - |\lambda_1(A)|^2 - |\lambda_2(A)|^2 =$$
$$a_{11}^2 + a_{12}^2 + a_{21}^2 + a_{22}^2 - 2[a_{11}a_{22} - a_{21}a_{12}].$$

Hence,

$$g(A) = \sqrt{(a_{11} - a_{22})^2 + (a_{21} + a_{12})^2}. \qquad (6.1)$$

Let $n = 2$ and a matrix A have real entries again, but now the *eigenvalues of A are real.* Then

$$|\lambda_1(A)|^2 + |\lambda_2(A)|^2 = Trace\ A^2.$$

Obviously,

$$A^2 = \begin{pmatrix} a_{11}^2 + a_{12}a_{21} & a_{11}a_{12} + a_{12}a_{22} \\ a_{21}a_{11} + a_{21}a_{22} & a_{22}^2 + a_{21}a_{12} \end{pmatrix}.$$

We thus get the relation

$$|\lambda_1(A)|^2 + |\lambda_2(A)|^2 = a_{11}^2 + 2a_{12}a_{21} + a_{22}^2.$$

Consequently,

$$g^2(A) = N^2(A) - |\lambda_1(A)|^2 - |\lambda_2(A)|^2 =$$
$$a_{11}^2 + a_{12}^2 + a_{21}^2 + a_{22}^2 - (a_{11}^2 + 2a_{12}a_{21} + a_{22}^2).$$

Hence,

$$g(A) = |a_{12} - a_{21}|. \qquad (6.2)$$

Example 1.6.2 *Let A be an upper-triangular matrix:*

$$A = \begin{pmatrix} a_{11} & a_{12} & \dots & a_{1n} \\ 0 & a_{21} & \dots & a_{2n} \\ . & \dots & . & . \\ 0 & \dots & 0 & a_{nn} \end{pmatrix}.$$

Then

$$g(A) = \sqrt{\sum_{k=1}^{n}\sum_{j=1}^{k-1} |a_{jk}|^2}, \qquad (6.3)$$

since the eigenvalues of a triangular matrix are its diagonal elements.

Example 1.6.3 *Consider the matrix*

$$A = \begin{pmatrix} -a_1 & \dots & -a_{n-1} & -a_n \\ 1 & \dots & 0 & 0 \\ . & \dots & . & . \\ 0 & \dots & 1 & 0 \end{pmatrix}$$

with complex numbers a_k. Such matrices play a key role in the theory of scalar ordinary differential equations. Take into account that

$$A^2 = \begin{pmatrix} a_1^2 - a_2 & \dots & a_1 a_{n-1} - a_1 & a_1 a_n \\ -a_1 & \dots & -a_{n-1} & -a_n \\ 1 & \dots & 0 & 0 \\ . & \dots & . & . \\ 0 & \dots & 0 & 0 \end{pmatrix}.$$

Thus, we obtain, $Trace\ A^2 = a_1^2 - 2a_2$. Therefore

$$g^2(A) \le N^2(A) - |Trace\ A^2| = n - 1 - |a_1^2 - 2a_2| + \sum_{k=1}^{n} |a_k|^2. \tag{6.4}$$

Hence,

$$g^2(A) \le n - 1 + 2|a_2| + \sum_{k=2}^{n} |a_k|^2. \tag{6.5}$$

1.7 Integral Inequalities

Let $h = (h_k)_{k=1}^n$ and $w = (w_k)_{k=1}^n$ be real vectors. We will write

$$h \le w \text{ if } w_k \le h_k \ (k = 1, ..., n).$$

The vector w is non-negative, if

$$w_k \ge 0 \ (k = 1, ..., n).$$

A vector-valued function is non-negative if its values are non-negative vectors. In the sequel we will often use the following result.

Lemma 1.7.1 *Let a non-negative continuous vector-valued function* $\phi : R_+ \to \mathbf{R}^n$ *satisfy the inequality*

$$\phi(t) \le f(t) + \int_0^t K(t,s)\phi(s)ds \ \ (t \ge 0)$$

with a non-negative continuous vector-valued function $f : R_+ \to \mathbf{R}^n$ *and a continuous matrix-valued kernel* $K : R_+^2 \to \mathbf{R}^n$ *with non-negative entries. Then* $\phi(t) \le \psi(t)$*, where* $\psi(t)$ *is a solution of the integral equation*

$$\psi(t) = f(t) + \int_0^t K(t,s)\psi(s)ds.$$

Similarly, the inequality

$$\phi(t) \geq f(t) + \int_0^t K(t,s)\phi(s)ds \quad (t \geq 0)$$

implies $\phi(t) \geq \psi(t) \quad (t \geq 0)$.

Proof: The lemma is due to an obvious application of the well-known Theorem 1.9.3 (Daleckii and Krein, 1974) to the space $C(R_+, \mathbf{R}^n)$ of real continuous vector-valued functions defined on R_+, since the integral operator

$$(Ku)(t) = \int_0^t K(t,s)u(s)ds$$

is a Volterra operator and, as is well known (Gohberg and Krein, 1970), its spectral radius equals zero. □

Corollary 1.7.2 *(The Gronwall inequality). Suppose a positive continuous scalar-valued function* ϕ *subordinates the inequality*

$$\phi(t) \leq c + \int_0^t h(s)\phi(s)ds \ (c = const > 0, \ t \geq 0),$$

where $h(t)$ *is a continuous positive scalar-valued function. Then*

$$\phi(t) \leq c \, exp\, [\int_0^t h(s)ds] \quad (t \geq 0).$$

Corollary 1.7.3 *Suppose a positive continuous scalar-valued function* ϕ *subordinates the inequality*

$$\phi(t) \leq c \, exp\, [\int_0^t p(\tau)d\tau] + \int_0^t \, exp\, [\int_s^t p(\tau)d\tau]h(s)\phi(s)ds \quad (t \geq 0),$$

where $h(t)$ *and* $p(t)$ *are continuous scalar-valued functions. Moreover,* $h(t)$ *is positive. Then*

$$\phi(t) \leq c \, exp\, [\int_0^t (p(s) + h(s))ds] \quad (t \geq 0).$$

To establish that result it is sufficient to set

$$\phi_1(t) = \phi(t) \, exp\, [-\int_0^t p(s)ds]$$

and to apply Corollary 1.7.2.

1.8 Algebraic Equations

Let us consider the algebraic equation

$$z^n = P(z), \tag{8.1}$$

where $P(z)$ is the polynomial

$$P(z) = \sum_{j=0}^{n-1} c_j z^{n-j-1}$$

with non-negative coefficients c_j $(j = 0, ..., n-1)$.

Lemma 1.8.1 *The extreme right-hand (non-negative) root z_0 of equation (8.1) satisfies the estimate $z_0 \leq \delta_0$, where*

$$\delta_0 = \begin{cases} [P(1)]^{1/n} & \text{if } P(1) \leq 1 \\ P(1) & \text{if } P(1) > 1 \end{cases}$$

For the proof see (Gil', 2003, Section 1.6).

Setting in (8.1) $z = ax$ with a positive constant a, we obtain

$$x^n = \sum_{j=0}^{n-1} c_j a^{-j-1} x^{n-j-1}. \tag{8.2}$$

If

$$a = 2 \max_{j=0,...,n-1} \sqrt[j+1]{c_j},$$

then

$$\sum_{j=0}^{n-1} c_j a^{-j-1} \leq \sum_{j=0}^{n-1} 2^{-j-1} = 1 - 2^{-n+1} < 1.$$

Let x_0 be the extreme right-hand root of equation (8.2), then by the previous lemma $x_0 \leq 1$. Since $z_0 = ax_0$, we have derived

Corollary 1.8.2 *The extreme right-hand root z_0 of equation (8.1) is non-negative. Moreover,*

$$z_0 \leq 2 \max_{j=0,...,n-1} \sqrt[j+1]{c_j}.$$

1.9 Upper Bounds for Lyapunov's Equation

Recall the famous Lyapunov theorem

Theorem 1.9.1 *In order for the eigenvalues of a matrix A to lie in the interior of the left half-plane, it is necessary and sufficient that there exists a positive definite Hermitian matrix W, such that the matrix $WA + A^*W$ is a negative Hermitian one. Moreover, if the eigenvalues of A lie in the interior of the left half-plane, then for any positive definite Hermitian matrix H there exists a positive definite Hermitian matrix W_H such that*

$$W_H A + A^* W_H = -2H. \tag{9.1}$$

In addition,

$$W_H = 2\int_0^\infty e^{A^*t} H e^{At} dt. \tag{9.2}$$

For the proof see for instance, (Daleckii and Krein, 1974, p. 33) or (Vidyasagar, 1993, p. 198). Moreover,

$$W_H = \frac{1}{\pi}\int_{-\infty}^{\infty} (-iI\omega - A_0^*)^{-1} H (iI\omega - A_0)^{-1} d\omega, \tag{9.3}$$

cf. (Godunov, 1998, p.151).

Equation (9.1) is called *the Lyapunov equation.* In many applications, it is important to know bounds for the norm of the solution W_H of the Lyapunov equation. We are beginning with

Lemma 1.9.2 *Let A be a Hurwitz $n \times n$-matrix. Then a solution W_H of the Lyapunov equation (9.1) subordinates the inequality*

$$\|W_H\| \le \|H\| \sum_{j,k=0}^{n-1} \frac{g^{j+k}(A)(k+j)!}{2^{j+k}|\alpha(A)|^{j+k+1}(j!\ k!)^{3/2}}.$$

Proof: Take into account that $g(A) = g(A^*)$, and $\alpha(A) = \alpha(A^*)$. Then by virtue of Corollary 1.5.3 and equality (9.2),

$$\|W_H\| \le 2\int_0^\infty \|e^{A^*t}\| \|H\| \|e^{At}\| dt \le$$

$$2\|H\| \int_0^\infty \exp[2\alpha(A)t] \Big(\sum_{k=0}^{n-1} \frac{g^k(A)t^k}{(k!)^{3/2}}\Big)^2 dt.$$

The integration gives us

$$\|W_H\| \le 2\|H\| \int_0^\infty \exp[2\alpha(A)t] \sum_{j,k=0}^{n-1} \frac{(g(A)t)^{k+j}}{(j!\ k!)^{3/2}} dt =$$

$$\|H\| \sum_{j,k=0}^{n-1} 2\frac{(k+j)!g^{j+k}(A)}{(2|\alpha(A)|)^{j+k+1}(j!\ k!)^{3/2}},$$

as claimed. □

Lemma 1.9.3 *Let A be a real Hurwitz $n \times n$-matrix. Then a solution W_H of the Lyapunov equation (9.1) subordinates the inequality*

$$\|W_H\| \le \|H\| \frac{1}{\pi} \int_{-\infty}^{\infty} \frac{(N^2(A) + n\omega^2)^{n-1} d\omega}{|det\ (i\omega I - A)|^2\ \ (n-1)^{n-1}}.$$

Proof: Since A is real and $Im\ Trace\ A(t) = 0$. Due to Lemma 1.5.7 , we have

$$\|(Iiy - A)^{-1}\| \le \frac{(N^2(A) + ny^2)^{(n-1)/2}}{|det\ (iyI - A)|\ \ (n-1)^{(n-1)/2}} \quad (y \in \mathbf{R}).$$

Hence, the required result is due to (9.3). □

Note that, if the spectrum is real, then

$$|det\ (i\omega I - A)|^2 \ge (\omega^2 + \alpha^2(A))^n \quad (\omega \in \mathbf{R}).$$

Thus, $\|W_H\| \le \|H\|\psi$, where

$$\psi := \frac{2\beta_n}{\pi} \int_0^{\infty} \frac{(N^2(A) + ny^2)^{n-1}}{(y^2 + \alpha^2(A))^n} dy,$$

where

$$\beta_n = \frac{1}{(n-1)^{n-1}}.$$

Clearly,

$$\psi = \frac{2\beta_n}{\pi|\alpha(A)|^{2n-1}} \int_0^{\infty} \frac{(N^2(A) + n\alpha^2(A)s^2)^{n-1}}{(s^2+1)^n} ds.$$

Take into account that

$$\sum_{k=1}^{n} |\lambda_k(A|^2 \le N^2(A)$$

and

$$|\alpha(A)| \le \min_{k=1,\dots,n} |\lambda_k(A)|.$$

So $n\alpha^2(A) \le N^2(A)$ and

$$\psi \le \frac{2^n \beta_n N^{2(n-1)}(A)}{\pi|\alpha(A)|^{2n-1}} \int_0^{\infty} \frac{ds}{s^2+1}.$$

Hence, $\psi \le \psi_1(A)$, where

$$\psi_1(A) = \frac{2^{n-1} N^{2(n-1)}(A)}{(n-1)^{n-1}|\alpha(A)|^{2n-1}}.$$

Now lemma 1.9.3 implies

Lemma 1.9.4 *Let A be a real Hurwitz $n \times n$-matrix with the real spectrum. Then a solution W_H of the Lyapunov equation (9.1) subordinates the inequality*

$$\|W_H\| \leq \|H\| \psi_1(A).$$

Let us consider the Lyapunov equation with $H = I$:

$$WA + A^*W = -2I. \tag{9.4}$$

Furthermore, let $u(t)$ be a solution of the equation

$$\dot{u}(t) = Au(t), \tag{9.5}$$

and let W be a solution of equation (9.4). Multiplying equation (9.5) by W and doing the scalar product, we get

$$(W\dot{u}(t), u(t)) = (WAu(t), u(t)).$$

Since

$$\frac{d}{dt}(Wu(t), u(t)) = (W\dot{u}(t), u(t)) + (u(t), W\dot{u}(t)),$$

it can be written

$$\frac{d}{dt}(Wu(t), u(t)) = (WAu(t), u(t)) + (u(t), WAu(t)) =$$

$$((WA + A^*W)u(t), u(t)).$$

Now (9.5) yields

$$\frac{d}{dt}(Wu(t), u(t)) = -2(u(t), u(t)) = -2(W^{-1}Wu(t), u(t)) \leq$$

$$-2b(W)(Wu(t), u(t)),$$

where

$$b(W) = \inf_{h \in \mathbf{C}^n} \frac{(W^{-1}h, h)}{(h, h)} = \inf_{w \in \mathbf{C}^n} \frac{(w, w)}{(Ww, w)} = \|W\|^{-1}.$$

Solving this inequality, we get

$$(Wu(t), u(t)) = \|W^{1/2}u(t)\|^2 \leq e^{-t2\|W\|^{-1}} \|W^{1/2}u(0)\|^2.$$

Since $u(t) = e^{At}u(0)$, this inequality means that

$$\|W^{1/2}e^{At}h\| \leq \|W^{1/2}h\| \; exp\,[-\frac{t}{\|W\|}] \;\; (h \in \mathbf{C}^n; \; t \geq 0). \tag{9.6}$$

We thus have derived.

Lemma 1.9.5 *Let A be a Hurwitz matrix, and W be a solution of equation (9.4). Then inequality (9.6) is true.*

This lemma, and Lemmas 1.9.2 and 1.9.4 yield

Corollary 1.9.6 *Let A be a real Hurwitz $n \times n$-matrix, and W be a solution of equation (9.4). Then with the notation*

$$\psi_0(A) := \sum_{j,k=0}^{n-1} \frac{g^{j+k}(A)(k+j)!}{2^{j+k}|\alpha(A)|^{j+k+1}(j!\ k!)^{3/2}},$$

the inequalities

$$\|W^{1/2}e^{At}\| \leq \sqrt{\psi_0(A)}\ e^{-t\psi_0^{-1}(A)} \text{ and}$$

$$\|W^{1/2}e^{At}\| \leq \sqrt{\psi_1(A)}\ e^{-t\psi_1^{-1}(A)} \quad (t \geq 0)$$

are true.

1.10 Lower Bounds for Lyapunov's Equation

Lemma 1.10.1 *Let A be a real Hurwitz $n \times n$-matrix. Then a solution W of equation (9.4) satisfies the inequality*

$$(Wh, h) \geq \frac{\|h\|^2}{\pi\|A\|} \quad (h \in \mathbf{C}^n).$$

Proof: By the Parseval equality we have

$$(Wh, h) = \int_0^\infty (e^{As}h, e^{As}h)ds$$

$$= \frac{1}{2\pi} \int_{-\infty}^\infty \|(A - iIy)^{-1}h\|^2\, dy \quad (h \in \mathbf{C}^n).$$

But

$$\|(A - iIy)^{-1}h\| \geq \frac{\|h\|}{\|A - iIy\|} \geq \frac{\|h\|}{\|A\| + |y|}.$$

Thus,

$$(Wh, h) \geq \frac{1}{2\pi} \int_{-\infty}^\infty \frac{\|h\|^2 dy}{(\|A\| + |y|)^2} =$$

$$\frac{\|h\|^2}{\pi} \int_0^\infty \frac{dy}{(\|A\| + |y|)^2} dy = \frac{\|h\|^2}{\pi\|A\|},$$

as claimed. □

Lemma 1.10.2 *Let A be a Hurwitz $n \times n$-matrix. Then a solution W_H of the Lyapunov equation (9.1) satisfies the inequality*

$$(W_H h, h) \geq 2\tilde{\theta}(H) \int_0^\infty e^{2\beta(A)t} \, [\sum_{k=0}^{n-1} \frac{g^k(A)t^k}{(k!)^{3/2}} \,]^{-2} dt \; \|h\|^2 \quad (h \in \mathbf{C}^n),$$

where

$$\beta(A) = \min_k Re\lambda_k(A), \tilde{\theta}(H) = \min_k \lambda_k(H).$$

Proof: Take into account that

$$g(A) = g(A^*) \text{ and } \beta(A) = \beta(A^*).$$

Then by Corollary 1.5.4,

$$(e^{A*t} H e^{At} h, h) = (He^{At} h, e^{At} h) \geq$$

$$\tilde{\theta}(H)(e^{At} h, e^{At} h) \geq$$

$$\tilde{\theta}(H) e^{2\beta(A)t} \, [\sum_{k=0}^{n-1} g^k(A) t^k (k!)^{-3/2}]^{-2} \; \|h\|^2 \quad (t \geq 0).$$

Now equality (9.2) yields

$$(W_H h, h) = 2 \int_0^\infty (e^{A*t} H e^{At} h, h) dt \geq$$

$$2\tilde{\theta}(H) \int_0^\infty e^{2\beta(A)t} \, [\sum_{k=0}^{n-1} g^k(A) t^k (k!)^{-3/2}]^{-2} dt \|h\|^2,$$

as claimed. □

If A is a normal matrix, then $g(A) = 0$ and the previous lemma gives us the inequality

$$(W_H h, h) \geq \frac{\tilde{\theta}(H)\|h\|^2}{|\beta(A)|} \quad (h \in \mathbf{C}^n).$$

1.11 Estimates for Contour Integrals

Lemma 1.11.1 *Let D be the closed convex hull of points $x_0, x_1, ..., x_n \in \mathbf{C}$ and let a scalar-valued function f be regular on a neighborhood D_1 of D. In addition, let $\Gamma \subset D_1$ be a Jordan closed contour surrounding the points $x_0, x_1, ..., x_n$. Then the inequality*

$$|\frac{1}{2\pi i} \int_\Gamma \frac{f(\lambda) d\lambda}{(\lambda - x_0)...(\lambda - x_n)}| \leq \frac{1}{n!} \sup_{\lambda \in D} |f^{(n)}(\lambda)|$$

is valid.

Proof: First, let all the points be distinct: $x_j \neq x_k$ for $j \neq k$ $(j,k = 0,...,n)$, and let $D_f(x_0, x_1, ..., x_n)$ be a divided difference of the scalar-valued function f at points $x_0, x_1, ..., x_n$ of the complex plane. If f is regular on a neighborhood of the closed convex hull D of the points $x_0, x_1, ..., x_n$, then the divided difference admits the representation

$$D_f(x_0, x_1, ..., x_n) = \frac{1}{2\pi i} \int_\Gamma \frac{f(\lambda)d\lambda}{(\lambda - x_0)...(\lambda - x_n)} \tag{11.1}$$

(see (Gelfond, 1967, formula (54)). But, on the other hand, the following estimate is well known:

$$| D_f(x_0, x_1, ..., x_n) | \leq \frac{1}{n!} \sup_{\lambda \in D} |f^{(n)}(\lambda)|$$

(Gelfond, 1967, formula (49)). Combining that inequality with relation (11.1), we arrive at the required result. If $x_j = x_k$ for some $j \neq k$, then the claimed inequality can be obtained by small perturbations and the previous reasonings. □

Lemma 1.11.2 *Let $x_0 \leq x_1 \leq ... \leq x_n$ be real points and let a function f be regular on a neighborhood D_1 of the segment $[x_0, x_n]$. In addition, let $\Gamma \subset D_1$ be a Jordan closed contour surrounding $[x_0, x_n]$. Then there is a point $\eta \in [x_0, x_n]$, such that the equality*

$$\frac{1}{2\pi i} \int_\Gamma \frac{f(\lambda)d\lambda}{(\lambda - x_0)...(\lambda - x_n)} = \frac{1}{n!} f^{(n)}(\eta)$$

is true.

Proof: First suppose that all the points are distinct: $x_0 < x_1 < ... < x_n$. Then the divided difference $D_f(x_0, x_1, ..., x_n)$ of f in the points $x_0, x_1, ..., x_n$ admits the representation

$$D_f(x_0, x_1, ..., x_n) = \frac{1}{n!} f^{(n)}(\eta)$$

with some point $\eta \in [x_0, x_n]$ (Gelfond, 1967, formula (43)), (Ostrowski, 1973, p.5). Combining that equality with representation (11.1), we arrive at the required result. If $x_j = x_k$ for some $j \neq k$, then the claimed inequality can be obtained by small perturbations and the previous reasonings. □

1.12 Absolute Values of Matrix Functions

1.12.1 Statement of the Result

Let $A = (a_{lj})_{j,l=1}^n$ be a matrix, and let $m_{lj}(\lambda)$ *be the cofactor of the element* $\delta_{lj}\lambda - a_{lj}$ *of the matrix* $\lambda I - A$. Here δ_{lj} is the Kronecker symbol: $\delta_{lj} = 0$ if

$j \neq l$, and $\delta_{jj} = 1$. Assume that after division by the coinciding zeros, the equality

$$\frac{m_{lj}(\lambda)}{det(I\lambda - A)} = \frac{\mu_{lj}(\lambda)}{d_{lj}(\lambda)} \quad (j, l = 1, ..., n) \tag{12.1}$$

holds, where $d_{lj}(\lambda)$ and $\mu_{lj}(\lambda)$ are polynomials which have no coinciding zeros. That is, $\mu_{lj}(\lambda)/d_{lj}(\lambda)$ are the entries of the matrix $(I\lambda - A)^{-1}$. Furthermore, let the degree of $d_{lj}(\lambda)$ be denoted by n_{lj}. Obviously, if $m_{lj}(\lambda)$ and $det(I\lambda - A)$ have no coinciding zeros, then $n_{lj} = n$ and

$$m_{lj}(\lambda) = \mu_{lj}(\lambda).$$

For $k = 0, ..., n-1$, put

$$b_{jl}(k) = \frac{1}{k!(n_{lj} - 1 - k)!} \sup_{\lambda \in co(A)} |\mu_{lj}^{(n_{lj}-k-1)}(\lambda)| \,.$$

Since $(n_{lj} - j)! = 0$ for $j > n_{lj}$, we have

$$b_{jl}(k) = 0 \text{ for } k = n_{lj}, ..., n-1 \text{ if } n_{lj} < n-1. \tag{12.2}$$

Theorem 1.12.1 *Let A be an $n \times n$-matrix, and let f be a function regular on a neighborhood of the closed convex hull $co(A)$ of the eigenvalues of A. Then the entries $f_{lj}(A)$ of the matrix $f(A)$ satisfy the inequalities*

$$|f_{lj}(A)| \leq \sum_{k=0}^{n-1} \max_{\lambda \in co(A)} |f^{(k)}(\lambda)| b_{lj}(k) \quad (l, j = 1, ..., n). \tag{12.3}$$

The proof of this theorem is presented in the next subsection.

Let $|A|$ denote the matrix of the absolute values of the elements of A and let $|h|$ have the similar interpretation for the vector h. Inequalities between vectors and between matrices are interpreted as inequalities between corresponding components. Then (12.3) can be rewritten in the form

$$|f(A)| \leq \sum_{k=0}^{n-1} \max_{\lambda \in co(A)} |f^{(k)}(\lambda)| B_k,$$

where

$$B_k = (b_{lj}(k))_{l,j=1}^{n} \quad (k = 0, 1, ..., n-1).$$

Corollary 1.12.2 *For any $n \times n$-matrix A we have the inequality*

$$|e^{At}| \leq exp[\alpha(A)t] \sum_{k=0}^{n-1} t^k B_k \quad (t \geq 0).$$

I.e. the entries $e_{lj}(At)$ of e^{At} satisfy the inequalities

$$|e_{lj}(At)| \leq exp[t\alpha(A)] \sum_{k=0}^{n-1} t^k b_{lj}(k) \quad (t \geq 0;\ l, j = 1, ..., n).$$

1.12.2 Proof of Theorem 1.12.1

Denote by $co_{lj}(A)$ the closed convex hull of the roots of the polynomial $d_{lj}(\lambda)$ defined by (12.1).

Lemma 1.12.3 *Let A be an $n \times n$-matrix, and let f be a function regular on a neighborhood of the closed convex hull $co(A)$ of the eigenvalues of A. Then the entries $f_{jl}(A)$ of $f(A)$ satisfy the inequalities*

$$|f_{jl}(A)| \leq \frac{1}{(n_{lj}-1)!} \sup_{\lambda \in co(A)} |\frac{d^{n_{lj}-1}(f(\lambda)\mu_{lj}(\lambda))}{d\lambda^{n_{lj}-1}}| \quad (l, j = 1, ..., n).$$

Proof: We begin with the representation

$$f(A) = \frac{1}{2\pi i} \int_{\Gamma} f(\lambda)(I\lambda - A)^{-1} d\lambda,$$

where Γ is a smooth contour containing all the eigenvalues of A. From the rule for inversion of matrices it follows that the entries $r_{jl}(\lambda)$ of the matrix $(I\lambda - A)^{-1}$ are

$$r_{jl}(\lambda) = \frac{m_{lj}(\lambda)}{det(I\lambda - A)} = \frac{\mu_{lj}(\lambda)}{d_{lj}(\lambda)} \quad (l, j = 1, ..., n).$$

Hence,

$$f_{jl}(A) = \frac{1}{2\pi i} \int_{\Gamma} \frac{f(\lambda)\mu_{lj}(\lambda)}{d_{lj}(\lambda)} d\lambda = \frac{1}{2\pi i} \int_{\Gamma} \frac{f(\lambda)\mu_{lj}(\lambda)}{(\lambda - \omega_1)...(\lambda - \omega_{n_{lj}})} d\lambda,$$

where $\omega_1, ..., \omega_{n_{lj}}$ are the roots of $d_{lj}(\lambda)$ which are simultaneously the eigenvalues of A taken with their multiplicities. Now by virtue of Lemma 1.11.1 we conclude that the required assertion is valid. □

Proof of Theorem 1.12.1: It can be written

$$|\frac{d^{n_{lj}-1}(f(\lambda)\mu_{lj}(\lambda))}{d\lambda^{n_{lj}-1}}| \leq \sum_{k=0}^{n_{lj}-1} C^k_{n_{lj}-1} |\mu_{lj}^{(n_{lj}-k-1)}(\lambda)||f^{(k)}(\lambda)|.$$

Recall that C^k_m are the binomial coefficients. Hence,

$$\sup_{\lambda \in co(A)} |\frac{d^{n_{lj}-1}(f(\lambda)\mu_{lj}(\lambda))}{d\lambda^{n_{lj}-1}}| \leq (n_{lj}-1)! \sum_{k=0}^{n_{lj}-1} b_{lj}(k) \sup_{\lambda \in co(A)} |f^{(k)}(\lambda)|.$$

Now Lemma 1.12.3 yields the required result. □

1.13 Banach Spaces

By $L^p \equiv L^p([0,\infty),\mathbf{C}^n)$ $(1 \le p < \infty)$ we denote the Banach space of functions defined on $[0,\infty)$ with values in $\mathbf{C}^n$ and the norm

$$\|f\|_{L^p} = [\int_0^\infty \|f(t)\|^p dt]^{1/p} \quad (f \in L^p)$$

and $C \equiv C([0,\infty),\mathbf{C}^n)$ is the Banach space of continuous functions defined on $[0,\infty)$ with values in $\mathbf{C}^n$ and the sup-norm norm

$$\|f\|_C = \sup_{t\ge 0} \|f(t)\| \quad (f \in C).$$

We will need the following

Lemma 1.13.1 *Let* $1/p + 1/q = 1$ $(p,q > 1)$. *Then for any function* $h \in L^p([0,\infty),\mathbf{C}^n)$ *with the property* $\dot{h} \in L^q([0,\infty),\mathbf{C}^n)$, *we have*

$$\|h(t)\|^2 \le 2[\int_t^\infty \|h(s)\|^p ds]^{1/p} \, [\int_t^\infty \|dh(s)/ds\|^q ds]^{1/q}.$$

Proof: Since $h(t)$ is a continuous function and $\|h\|_{L^p} < \infty$, due to the mean value theorem it can be written

$$\int_t^{t+1} \|h(s)\|^p ds = \|h(\theta)\|^p \to 0 \text{ as } t \to \infty \ (t \le \theta \le t+1).$$

Consequently, $h(\infty) = 0$. Obviously,

$$\|h(t)\|^2 = -\int_t^\infty \frac{d\|h(s)\|^2}{ds} ds = -2\int_t^\infty \|h(s)\| \frac{d\|h(s)\|}{ds} ds.$$

Taking into account that

$$|\frac{d}{ds}\|h(s)\|| \le \|dh(s)/ds\|,$$

we get the required result due to Hólder's inequality. □

2. Perturbations of Linear Systems

2.1 Evolution Operators

Let $u(t)$ be a solution of the equation

$$\dot{u}(t) = A(t)u(t) \quad (t \geq 0) \tag{1.1}$$

with a piecewise continuous $n \times n$-matrix $A(t)$. Then a linear operator (matrix)

$$U(t,s) : \mathbf{C}^n \to \mathbf{C}^n \quad (t, s \geq 0)$$

defined by the equality

$$U(t,s)u(s) = u(t)$$

is called *the evolution operator* of equation (1.1). The evolution operator has the following properties:

$$U(t,t) = I.$$

$$\frac{\partial}{\partial t}U(t,s) = A(t)U(t,s). \tag{1.2}$$

Moreover,

$$\frac{\partial}{\partial s}U(t,s) = -U(t,s)A(s) \tag{1.3}$$

and

$$U(t,s)U(s,\tau) = U(t,\tau) \quad (t, s, \tau \geq 0).$$

The latter property implies $U(s,t) = U^{-1}(t,s)$. If $U(t,s)$ is the evolution operator of equation (1.1), then $\tilde{U}(t,s) \equiv U^{-1}(t,s)$ is the evolution operator of the equation

$$\dot{u}(t) = -A(t)u(t) \; (t \geq 0). \tag{1.4}$$

The operator $U(t) = U(t,0)$ is called *the Cauchy operator* of equation (1.1). Since

$$U(t,s) = U(t,0)U(0,s), \text{ and } U(0,s) = U^{-1}(s,0),$$

the relation

$$U(t,s) = U(t)U^{-1}(s) \quad (t, s \geq 0)$$

is valid. If $A(t) \equiv A$ is a constant matrix, then

$$U(t) = e^{tA} \text{ and } U(t,s) = U(t-s) = e^{(t-s)A}.$$

Furthermore, a solution of the non-homogeneous equation

$$\dot{u}(t) = A(t)u(t) + f(t) \ (t \geq 0) \tag{1.5}$$

with a given piecewise continuous function $f : R_+ \to \mathbf{C}^n$ can be obtained in the form

$$u(t) = U(t,s)u(s) + \int_s^t U(t,\tau)f(\tau)d\tau \ \ (t \geq s \geq 0). \tag{1.6}$$

It is simple to check that, if $U(t,s)$ is the evolution operator of (1.1), then for any complex number α, $U_\alpha(t,s) = e^{\alpha(t-s)}U(t,s)$ is the evolution operator of the equation

$$\dot{u}(t) = (A(t) + \alpha I)u(t) \ \ (t \geq 0). \tag{1.7}$$

The Cauchy operator can be represented in the form

$$U(t) = I + \int_0^t A(t_1)dt_1 + \sum_{k=2}^{\infty} \int_0^t \int_0^{t_k} \dots \int_0^{t_2} A(t_k)\, A(t_{k-1}) \dots A(t_1) dt_k\, dt_{k-1} \dots dt_1.$$

Lemma 2.1.1 *A necessary and sufficient condition for the stability of equation (1.1) is the uniform boundedness of its Cauchy operator:*

$$\sup_{t \geq 0} \|U(t)\| < \infty.$$

For the proof see (Daleckii and Krein, 1974, Lemma 3.3.1).

Rephrasing the conditions of the lemma, we will say that a necessary and sufficient condition for the stability of equation (1.1) is the existence of a constant $M > 0$ such that any solution $u(t)$ of equation (1.1) satisfies the estimate

$$\|u(t)\| \leq M\|u(0)\| \ \ (t \geq 0).$$

The least value M_0 of this constant is clearly given by the equality

$$M_0 = \sup_{t \geq 0} \|U(t)\|.$$

Linear equation (1.1) is uniformly stable if there exists a constant $M > 0$ such that any solution $u(t)$ of this equation satisfies the following estimate for all $t \geq s \geq 0$:

$$\|u(t)\| \leq M\|u(s)\|.$$

Since $u(t) = U(t,s)u(s)$, it can be shown that the uniform stability of the equation is equivalent to the condition

$$\sup_{t \geq s \geq 0} \|U(t,s)\| < \infty. \tag{1.8}$$

Note that if $A(t) \equiv A$ is a constant matrix, then the stability condition

$$\sup_{t \geq 0} \|e^{tA}\| < \infty$$

coincides with the uniform stability condition.

2.2 Perturbations of Evolution Operators

Lemma 2.2.1 *Let $U(t,s)$ be the evolution operator of equation (1.1) and $U_1(t,s)$ be the evolution operator of the equation*

$$\dot{u}(t) = A_1(t)u(t) \ \ (t \geq 0) \tag{2.1}$$

with a variable $n \times n$-matrix $A_1(t) \neq A(t)$. Then

$$U(t,s) - U_1(t,s) = \int_s^t U(t,\tau)(A(\tau) - A_1(\tau))U_1(\tau,s)d\tau \ \ (t \geq s \geq 0).$$

Proof: Subtract (2.1) from (1.1) and put $\delta(t) = u(t) - u_1(t)$, where $u(t)$ is a solution of (1.1), and $u_1(t)$ is a solution of (2.1) with the initial conditions $u_1(s) = u(s) = h$. Here h is a given vector. We get

$$\dot{\delta}(t) = A(t)\delta(t) + (A(t) - A_1(t))u_1(t).$$

Taking into account that $\delta(s) = 0$ and applying formula (1.6), we arrive at the equality

$$\delta(t) = \int_s^t U(t,\tau)(A(\tau) - A_1(\tau))u_1(\tau)d\tau.$$

But

$$u_1(\tau) = U_1(t,s)u(s) \text{ and } \delta(t) = (U(t,s) - U_1(t,s))u(s).$$

This proves the result, since $u(s) = h$ is an arbitrary vector. □

Furthermore, Lemma 2.2.1 implies,

$$\|U(t,s) - U_1(t,s)\| \leq \int_s^t \|U(t,\tau)\| \, \|A(\tau) - A_1(\tau)\| \, \|U_1(\tau,s)\|d\tau \ (0 \leq s < t).$$

Hence,

$$\|U_1(t,s)\| \leq \|U(t,s)\| + \int_s^t \|U(t,\tau)\| \, \|A(\tau) - A_1(\tau)\| \, \|U_1(\tau,s)\|d\tau$$
$$(0 \leq s < t). \tag{2.2}$$

Lemma 1.7.1 yields

Lemma 2.2.2 *Let $U(t,s)$ and $U_1(t,s)$ be the evolution operators of the equations (2.1) and (1.1), respectively. Then $\|U_1(t,s)\| \leq \eta_s(t)$, where $\eta_s(t)$ is the solution of the equation*

$$\eta_s(t) = \|U(t,s)\| + \int_s^t \|U(t,\tau)\|\|A(\tau) - A_1(\tau)\|\eta_s(\tau)d\tau \ \ (0 \leq s < t).$$

Corollary 2.2.3 *Let the evolution operator $U(t,s)$ of equation (1.1) satisfy the inequality*

$$\|U(t,s)\| \le Me^{-\nu(t-s)} \ (t \ge s \ge 0) \tag{2.3}$$

with positive constants M and ν. Then the evolution operator $U_1(t,s)$ of equation (2.1) subordinates the inequality

$$\|U_1(t,s)\| \le Me^{-\nu(t-s)} e^{M\int_s^t \|A(\tau)-A_1(\tau)\|d\tau} \quad (s \le t). \tag{2.4}$$

Moreover,

$$\|U(t,s) - U_1(t,s)\| \le Me^{-\nu(t-s)} (e^{M\int_s^t \|A(\tau)-A_1(\tau)\|d\tau} - 1). \tag{2.5}$$

Indeed, to prove inequality (2.4) we need, according to Lemma 2.2.2, only to check that the right part of that inequality is a solution of the equation

$$\eta(t) = Me^{-\nu(t-s)} + M\int_s^t e^{-\nu(t-\tau)}\|A(\tau) - A_1(\tau)\|\eta(\tau)d\tau \ (s \le t).$$

The checking of inequality (2.5) is left to the reader (see also (Daleckii and Krein, 1974, Section 3.2)). The latter corollary and relation (1.8) imply

Corollary 2.2.4 *Let equation (1.1) be uniformly exponentially stable and let*

$$\int_0^\infty \|A(\tau) - A_1(\tau)\|d\tau < \infty.$$

Then equation (2.1) is also uniformly exponentially stable.

If $A_1(t) = 0$, then by (2.5), we easily get:

Corollary 2.2.5 *Under condition (2.3) the evolution operator $U(t,s)$ of equation (1.1) satisfies the inequality*

$$\|U(t,s) - I\| \le Me^{-\nu(t-s)} (e^{\int_s^t \|A(\tau)\|d\tau} - 1) \ \ (s \le t).$$

Lemma 2.2.6 *Let $U(t,s)$ and $U_1(t,s)$ be the evolution operators of equations (1.1) and (2.1), respectively. For some positive $T \le \infty$, let the condition*

$$J \equiv \sup_{0\le t\le T} \int_0^t \|U(t,s)\|\|A(s) - A_1(s)\|ds < 1$$

be fulfilled. Then the inequality

$$\sup_{0\le t\le T} \|U_1(t,0)\| \le (1-J)^{-1} \sup_{0\le t\le T} \|U(t,0)\|$$

holds.

Proof: According to (2.2),

$$\sup_{0\le t\le T} \|U_1(t,0)\| \le \sup_{0\le t\le T} \|U(t,0)\| +$$

$$\sup_{0\le t\le T} \|U_1(t,0)\| \int_0^t \|U(t,\tau)\| \|A(\tau) - A_1(\tau)\| d\tau \le$$

$$\sup_{0\le t\le T} \|U(t,0)\| + \sup_{0\le \tau\le T} \|U_1(\tau,0)\| J.$$

Now the result is due to the condition $J < 1$. □

Lemma 2.2.7 *Let $U(t,s)$ and $U_1(t,s)$ be the evolution operators of equations (1.1) and (2.1), respectively. Let the condition*

$$\|U(t,s)\| \le c\, exp[\int_s^t a(\tau)]d\tau \ (0 \le s \le t)$$

be fulfilled with a continuous scalar-valued function $a(t)$ and a positive constant c. Then the inequality

$$\|U_1(t,s)\| \le c\, exp[\int_s^t (a(\tau) + c\|A(\tau) - A_1(\tau)\|) d\tau]$$

holds.

Proof: Thanks to Lemma 2.2.2, $\|U_1(t,0)\| \le \eta(t)$, where $\eta(t)$ is the solution of the equation

$$\eta(t) = c\, exp[\int_s^t a(\tau)d\tau] + c\int_s^t exp[\int_\tau^t a(\tau_1)d\tau_1]\|A(\tau) - A_1(\tau)\|\eta(\tau)d\tau.$$

Put

$$z(t) = \eta(t) exp[-\int_0^t a(\tau)d\tau].$$

Then

$$z(t) = c + c\int_s^t \|A(\tau) - A_1(\tau)\| z(\tau) d\tau.$$

Now the result is due to the Gronwall lemma. □

2.3 Representation of Solutions

For a bounded Riemann integrable matrix-valued function F defined on a segment [0,t], define *the left multiplicative integral* as the limit of the sequence of the products

$$\prod_{1\le k\le M-1}^{\leftarrow}(I+F(t_k^{(M)})\delta_k) :=$$

$$(I+F(t_{M-1}^{(M)})\delta_{M-1})(I+F(t_{M-2}^{(M)})\delta_{M-2})...(I+F(t_1^{(M)})\delta_1)$$

as $\max_k |t_{k+1}^{(M)} - t_k^{(M)}|$ tends to zero. Here

$$0 = t_1^{(M)} < t_2^{(M)} < ... < t_M^{(M)} = t$$

and

$$\delta_k = t_{k+1}^{(M)} - t_k^{(M)} \ (k = 1, ..., M-1).$$

I.e., the arrow over the symbol of the product means that the indexes of the co-factors increase from right to left. *The left multiplicative integral we denote by the symbol*

$$\int_{[0,t]}^{\leftarrow}(I+F(s)ds).$$

Lemma 2.3.1 *Any solution $x(t)$ of equation (1.1) with a piecewise continuous matrix $A(t)$ can be represented by the relation*

$$x(t) = \int_{[0,t]}^{\leftarrow}(I+A(s)ds)x(0).$$

For the proof see (Dollard and Friedman, 1979), (Gil', 1998, Section 4.1).

2.4 Triangular Systems

Let us consider the triangular system

$$\dot{u}_j(t) = \sum_{k=1}^{j} a_{jk}(t)u_k(t) \ \ (t \ge 0; \ j = 1, ..., n) \tag{4.1}$$

with piecewise continuous real functions $a_{jk}(t)$ $(1 \le k \le j \le n)$. Then we have

$$u_j(t) = e^{\int_0^t a_{jj}(s)ds}\, u_j(0) + \int_0^t e^{\int_\tau^t a_{jj}(s)ds}\sum_{k=1}^{j-1} a_{jk}(\tau)u_k(\tau)d\tau.$$

Hence we arrive at the inequality

$$|u_j(t)| \le e^{\int_0^t \alpha(s)ds}\, |u_j(0)| + \int_0^t e^{\int_\tau^t \alpha(s)ds}\sum_{k=1}^{j-1}|a_{jk}(\tau)||u_k(\tau)|d\tau$$

with the notation

$$\alpha(t) = \max_{j=1,\dots,n} a_{jj}(t). \tag{4.2}$$

Furthermore, putting

$$e^{-\int_0^t \alpha(s)ds}|u_j(t)| = w_j(t), \tag{4.3}$$

we get

$$|w_j(t)| \le |u_j(0)| + \int_0^t \sum_{k=1}^{j-1} |a_{jk}(\tau)||w_k(\tau)|d\tau \ (j = 1, ..., n). \tag{4.4}$$

Denote by $V(t)$ the lower-triangular matrix with the entries

$$v_{jk}(t) = |a_{jk}(t)| \text{ for } j > k, \text{ and } v_{jk}(t) = 0 \text{ for } j \le k \ (j, k = 1, ...n),$$

then we can write down inequalities (4.4) in the vector form

$$|w(t)| \le |u(0)| + \int_0^t V(\tau)|w(\tau)|d\tau.$$

Here $w(t) = (w_1(t), ..., w_n(t))$. For any continuous vector-valued function $h(t)$ defined on the positive half-line, introduce an operator W, defined by the relation

$$(Wh)(t) = \int_0^t V(\tau)h(\tau)d\tau.$$

But $V(t)$ is a lower-triangular $n \times n$-matrix with the zero diagonal. Therefore

$$V(\tau_1)V(\tau_2)...V(\tau_n) = 0 \ (\tau_1, ..., \tau_n \ge 0).$$

Consequently $W^n = 0$, and

$$(I - W)^{-1} = \sum_{k=0}^{\infty} W^k = \sum_{k=0}^{n-1} W^k. \tag{4.5}$$

It can be written as

$$|w(t)| \le (I - W)^{-1}|u(0)| = \sum_{k=0}^{n-1} W^k|u(0)|.$$

But

$$\|V(t)\| \le N(V(t)) = \sqrt{\sum_{1 \le k < j \le n} a_{jk}^2(t)}. \tag{4.6}$$

The latter inequality yields the relation

$$\|(W^k h)(t)\| \le \int_0^t \dots \int_0^{t_{k-1}} \|V(t_1) \dots V(t_k)\| dt_k \dots dt_1 \le$$

$$\int_0^t \dots \int_0^{t_{k-1}} N(V(t_1))\dots N(V(t_k))dt_k\dots dt_1 = \frac{1}{k!}[\int_0^t N(V(t_1))dt_1]^k.$$

This inequality and (4.4) imply

$$\|w(t)\| \leq \|(I-W)^{-1}\| \, \|u(0)\| \leq \|u(0)\| \sum_{k=0}^{n-1} \frac{1}{k!}[\int_0^t N(V(t_1))dt_1]^k.$$

According to (4.3), we arrive at the inequality

$$\|u(t)\| \leq \|u(0)\| e^{\int_0^t \alpha(s)ds} \sum_{k=0}^{n-1} \frac{1}{k!}[\int_0^t N(V(t_1))dt_1]^k.$$

Now instead of 0 let us take an arbitrary real τ. We thus have derived

Theorem 2.4.1 *With the notations (4.2) and (4.6) the inequality*

$$\|u(t)\| \leq \|u(\tau)\| e^{\int_\tau^t \alpha(s)ds} \sum_{k=0}^{n-1} \frac{1}{k!}[\int_\tau^t N(V(t_1))dt_1]^k \ (0 \leq \tau \leq t < \infty)$$

is true for any solution $u(t)$ of system (4.1).

Corollary 2.4.2 *Let the relation*

$$\overline{\lim}_{t\to\infty}\{\int_0^t \alpha(s)ds + \ln [\sum_{k=0}^{n-1} \frac{1}{k!} (\int_0^t N(V(t_1))dt_1)^k]\} < +\infty$$

hold. Then system (4.1) is stable.

Corollary 2.4.3 *Let the relation*

$$\overline{\lim}_{t\to\infty} t^{-1}\{\int_0^t \alpha(s)ds + \ln [\sum_{k=0}^{n-1} \frac{1}{k!} (\int_0^t N(V(t_1))dt_1)^k]\} < 0$$

hold. Then system (4.1) is exponentially stable.

2.5 Perturbations of Triangular Systems

Let $A(t) = (a_{jk}(t))_{j,k=1}^n$ be a real piecewise continuous $n \times n$-matrix. Denote the corresponding lower-triangular matrix by $T(t)$. That is, $T(t) = (T_{jk}(t))_{j,k=1}^n$, where

$$T_{jk}(t) = a_{jk}(t) \text{ if } j \geq k \text{ and } T_{jk}(t) = 0 \text{ if } j < k.$$

We will investigate system (1.1) as a perturbation of the triangular system

$$\dot{w}(t) = T(t)w. \tag{5.1}$$

Due to Theorem 2.4.1, a solution $w(t)$ of equation (5.1) is subject to the inequality

$$\|w(t)\| \leq e^{\int_\tau^t \alpha(s)ds} \sum_{k=0}^{n-1} \frac{1}{k!}[\int_\tau^t N(V(t_1))dt_1]^k \|w(\tau)\|. \tag{5.2}$$

Here $\alpha(t)$ is defined by (4.2) and $N(V(t))$ is defined by (4.6). Put

$$q(t) = \|A(t) - T(t)\|.$$

Employing Lemma 2.2.6 and inequality (5.2), we get

Theorem 2.5.1 *Let the condition*

$$W_A \equiv \sup_{t\geq 0} \int_0^t q(\tau) e^{\int_\tau^t \alpha(s)ds} \sum_{k=0}^{n-1} \frac{1}{k!}[\int_\tau^t N(V(t_1))dt_1]^k d\tau < 1$$

be fulfilled. Then equation (1.1) is stable. Moreover, the Cauchy operator $U(t)$ of equation (1.1) satisfies the inequality

$$\|U(t)\| \leq (1 - W_A)^{-1} \sup_{t\geq 0} e^{\int_0^t \alpha(s)ds} \sum_{k=0}^{n-1} \frac{1}{k!}[\int_0^t N(V(t_1))dt_1]^k.$$

2.6 Integrally Small Perturbations

Consider the equation

$$\dot{x} = A(t)x + B(t)x, \tag{6.1}$$

where $B(t)$ is a variable $n \times n$-matrix. Introduce the notations

$$J(t) = \int_0^t B(s)ds,$$

$$q(J) := \sup_{t\geq 0} \| \int_0^t B(s)ds\|,$$

and

$$m(t) = \|A(t)J(t) - J(t)(A(t) + B(t))\|.$$

Theorem 2.6.1 *Let $U(t, s)$ be the evolution operator of equation (1.1) and the condition*

$$\eta(J) \equiv q(J) + \sup_{t\geq 0} \int_0^t \|U(t, s)\| m(s)ds < 1 \tag{6.2}$$

hold. Then for any solution $x(t)$ of (6.1) the estimate

$$\|x(t)\| \leq \frac{\|x(0)\|}{1-\eta(J)} \max_{t\geq 0} \|U(t,0)\| \tag{6.3}$$

is true.

Proof: Rewrite (6.1) in the form

$$x(t) = U(t,0)x(0) + \int_0^t U(t,s)B(s)x(s)ds. \tag{6.4}$$

Evidently,

$$\int_0^t \frac{d}{ds}[U(t,s)J(s)x(s)]ds = J(t)x(t),$$

since $U(t,t)$ is the unit matrix. Exploiting equality (1.3), we obtain

$$J(t)x(t) = \int_0^t U(t,s)[-A(s)J(s)x(s)+ \\ B(s)x(s) + J(s)\dot{x}(s)]ds. \tag{6.5}$$

Substituting the right part of (6.1) instead of $\dot{x}(s)$ into (6.5), we arrive at the equality

$$\int_0^t U(t,s)B(s)x(s)ds = J(t)x(t) + \int_0^t U(t,s)[A(s)J(s)- \\ J(s)(A(s)+B(s))]x(s)ds.$$

Now (6.4) and the latter equality imply

$$x(t) = U(t,0)x(0) + J(t)x(t) + \int_0^t U(t,s)[A(s)J(s)- \\ J(s)(A(s)+B(s))]x(s)ds.$$

Hence,

$$\|x(t)\| \leq \|x(0)\|\|U(t,0)\| + q(J)\|x(t)\| + \int_0^t \|U(t,s)\|m(s)\|x(s)\|ds.$$

This inequality implies

$$y(t) \leq \|U(t,0)\| + q(J)y(t) + \int_0^t \|U(t,s)\|m(s)y(s)ds \leq \\ \|U(t,0)\| + \eta(J)\sup_{t\geq 0} y(t),$$

where

$$y(t) = \frac{\|x(t)\|}{\|x(0)\|}.$$

Now condition (6.2) ensures the required result. □

2.7 Integrally Small Perturbations of Autonomous Systems

Assume that $A(t) \equiv A$ is a constant matrix and consider the equation

$$dx/dt = Ax + B(t)x. \tag{7.1}$$

Suppose that

$$m_0 := \sup_{t\geq 0} \|AJ(t) - J(t)(A + B(t))\| < \infty$$

and

$$q(J) + m_0 \sum_{k=0}^{n-1} \frac{g^k(A)}{\sqrt{k!}|\alpha(A)|^{k+1}} < 1. \tag{7.2}$$

Recall that $g(A)$ is defined in Section 1.5.

Theorem 2.7.1 *Let condition (7.2) hold. Then system (7.1) is stable.*

Proof: In the case of a constant matrix the evolution operator $U(t,s)$ of the equation (1.1) is equal to $exp[A(t-s)]$. By Corollary 1.5.3 we easily get

$$\int_0^t \|exp[A(t-s)]\|ds \leq \int_0^\infty \|exp[At]\|ds \leq$$

$$\sum_{k=0}^{n-1} \frac{g^k(A)}{\sqrt{k!}|\alpha(A)|^{k+1}}.$$

Now Theorem 2.6.1 implies the stated result □ .

Let $C(t)$ be a *τ-periodic matrix.* Take

$$A = \tau^{-1} \int_0^\tau C(t)dt, \; B(t) = C(t) - A. \tag{7.3}$$

Besides,

$$J(\tau) = \int_0^\tau B(t)dt = 0,$$

$$q(J) = \sup_{0\leq t\leq\tau} \|J(t)\|, \tag{7.4}$$

and

$$m_0 = \sup_{0\leq t\leq\tau} \|AJ(t) - J(t)(A + B(t))\|. \tag{7.5}$$

The periodic system

$$\dot{x}(t) = C(t)x \tag{7.6}$$

can be rewritten in the shape (7.1). Now Theorem 2.7.1 gives us the stability condition for system (7.6).

Example 2.7.2 *Consider system (7.6) wih real τ-periodic matrix $C(t) = (c_{jk}(t))_{j,k=1}^2$. That is $n = 2$.*

Define $A, B(t), q(J)$ and m_0 by (7.3)-(7.5). That is,

$$a_{jk} = \tau^{-1} \int_0^\tau c_{jk}(t)dt, \ B(t) = C(t) - A.$$

According to Example 1.6.1,

$$g(A) \leq |a_{12} - a_{21}|.$$

Thus according to Theorem 2.7.1, the condition

$$q(J) + \frac{m_0}{|\alpha(A)|}(1 + \frac{|a_{12} - a_{21}|}{|\alpha(A)|}) < 1$$

provides the stability of the considered periodic system.

3. Systems with Slowly Varying Coefficients

3.1 The Freezing Method

Consider in $\mathbf{C}^n$ the equation

$$\dot{x}(t) = A(t)x(t) \quad (t \geq 0), \tag{1.1}$$

where $A(t)$ is a variable piecewise-continuous Hurwitz $n \times n$ - matrix. For a fixed $s \geq 0$, let

$$\|exp[A(s)t]\| \leq p(t, A(s)) \quad (t, s \geq 0), \tag{1.2}$$

where $p(t, A(s))$ is a piecewise-continuous in t scalar function. Denote

$$q(t, s) := \|A(t) - A(s)\| \quad (t, s \geq 0),$$

and assume that

$$\chi := \sup_{t \geq 0} p(t, A(t)) < \infty. \tag{1.3}$$

Theorem 3.1.1 *Let the conditions (1.2), (1.3) and*

$$\theta := \sup_{t \geq 0} \int_0^t p(t - t_1, A(t))q(t, t_1)dt_1 < 1$$

hold. Then equation (1.1) is uniformly stable. Moreover, the estimate

$$\|x(t)\| \leq \frac{\chi\|x(s)\|}{1 - \theta} \quad (t > s \geq 0) \tag{1.4}$$

is true for any solution $x(t)$ of (1.1).

The proofs of this theorem and the next one are presented in the next section.

Corollary 3.1.2 *Let*

$$\|exp[A(s)t]\| \leq p_1(t) \quad (t, s \geq 0),$$

where $p_1(t)$ is a piecewise-continuous function independent of s. In addition, let

$$\|A(t) - A(s)\| \leq q_0|t - s| \quad (t, s \geq 0;\ q_0 \equiv const > 0), \tag{1.5}$$

and

$$\theta_0 := q_0 \int_0^\infty t p_1(t) dt < 1. \tag{1.6}$$

Then equation (1.1) is uniformly stable. Moreover, estimate (1.4) is true with $\theta = \theta_0$ *and* $\chi = \chi_0$, *where*

$$\chi_0 := \sup_{t \ge 0} p_0(t).$$

Recall that the quantity $g(A)$ is defined in Section 1.5. In addition, for any $n \times n$-matrix A Corollary 1.5.3 gives us the inequality:

$$\|exp[A(s)t]\| \le \tilde{p}(t, A(s)) \ \ (t, s \ge 0), \tag{1.7}$$

where

$$\tilde{p}(t, A(s)) := exp[\alpha(A(s))t] \sum_{k=0}^{n-1} \frac{g^k(A(s))t^k}{(k!)^{3/2}}$$

with $\alpha(A(s)) = \max_k Re\ \lambda_k(A(s))$. Let

$$\tilde{\chi} := \sup_{t \ge 0} \tilde{p}(t, A(t)) < \infty.$$

Now Theorem 3.1.1 implies

Corollary 3.1.3 *Let the condition*

$$\tilde{\zeta} := \sup_{t \ge 0} \int_0^t \tilde{p}(t - t_1, A(t)) q(t, t_1) dt_1 < 1$$

hold. Then equation (1.1) is stable. Moreover, the estimate

$$\|x(t)\| \le \frac{\tilde{\chi}\|x(s)\|}{1 - \tilde{\zeta}} \ \ (t > s \ge 0)$$

is true for any solution $x(t)$ *of (1.1).*

Let

$$v := \sup_{t \ge 0} g(A(t)) < \infty \tag{1.8}$$

and

$$\rho := -\sup_{t \ge 0} \alpha(A(t)) > 0. \tag{1.9}$$

Denote

$$\tilde{p}_0(t) := \exp[-\rho t] \sum_{k=0}^{n-1} \frac{t^k v^k}{(k!)^{3/2}},$$

and

$$\tilde{\chi}_0 := \sup_{t \ge 0} \tilde{p}_0(t).$$

Simple calculations show that

$$\int_0^\infty t\tilde{p}_0(t)dt = \zeta_1,$$

where

$$\zeta_1 := \sum_{k=0}^{n-1} \frac{(k+1)v^k}{\sqrt{k!}\rho^{2+k}}.$$

Now Corollary 3.1.2 and a small perturbation imply

Corollary 3.1.4 *Let conditions (1.5), (1.8) and (1.9) hold. In addition, let* $q_0\zeta_1 < 1$. *Then equation (1.1) is exponentially stable. Moreover, the evolution operator* $U(t,s)$ *of equation (1.1) satisfies the inequality*

$$\|U(t,s)\| \le \frac{\tilde{\chi}_0}{1-q_0\zeta_1} \quad (t,s \ge 0).$$

Furthermore, denote by $z(q_0, v)$ the extreme right-hand root of the algebraic equation

$$z^{n+1} = q_0 P(z), \tag{1.10}$$

where

$$P(z) = \sum_{k=0}^{n-1} \frac{(k+1)v^k}{\sqrt{k!}} z^{n-k-1}.$$

Recall that I is the unit matrix.

Theorem 3.1.5 *Let conditions (1.5) and (1.8) hold. In addition, let the matrix*

$$A(t) + (\epsilon + z(q_0, v))I$$

be a Hurwitz one for an $\epsilon > 0$ *and all* $t \ge 0$. *Then equation (1.1) is exponentially stable. Moreover, there is a positive constant* a, *such that any solution* $x(t)$ *of equation (1.1) satisfies the inequalities*

$$a^{-1} exp[(\beta_0 - z(q_0,v))t] \le \frac{\|x(t)\|}{\|x(0)\|} \le a\ exp[(-\rho + z(q_0,v))t] \ (t \ge 0),$$

where

$$\beta_0 = \inf_{t\ge 0} \min_k Re\lambda_k(A(t)).$$

Setting $z = vy$ in (1.10), we have the equation

$$y^{n+1} = q_0 v^{-2} \sum_{k=0}^{n-1} \frac{k+1}{\sqrt{k!}} y^{n-k-1}.$$

Hence due to Lemma 1.8.1, with the notation

$$w_n = \sum_{k=0}^{n-1} \frac{k+1}{\sqrt{k!}},$$

we get $z(q_0, v) \leq \delta(q_0, v)$, where

$$\delta(q_0, v) = \begin{cases} v^{1-\frac{2}{n+1}} [q_0 w_n]^{1/(n+1)} & \text{if } q_0 w_n \geq v^2, \\ q_0/v & \text{if } q_0 w_n \leq v^2. \end{cases}$$

Theorem 3.1.5 yields

Corollary 3.1.6 *Let conditions (1.5) and (1.8) hold. If, in addition, the matrix*

$$A(t) + (\epsilon + \delta(q_0, v))I$$

is a Hurwitz one for an $\epsilon > 0$ and all $t \geq 0$, then system (1.1) is exponentially stable.

Below we also will prove the following

Lemma 3.1.7 *Let conditions (1.5), (1.8) and (1.9) be fulfilled. In addition, let*

$$q_0 < \frac{\rho^{n+1}}{P(\rho)}. \tag{1.11}$$

Then the matrix $A(t) + z(q_0, v)I$ is a Hurwitz one.

Thus under (1.5), (1.8), (1.9) and (1.11) system (1.1) is stable.

Example 3.1.8 *Let us consider the equation*

$$\ddot{y} + a(t)\dot{y} + b(t)y = 0 \ (t \geq 0).$$

Here $a(t)$ and $b(t)$ are bounded on $[0, \infty)$ positive scalar-valued functions with the property

$$|a(t) - a(s)| + |b(t) - b(s)| \leq q_0|t - s| \ \ (t, s \geq 0).$$

According to Example 1.6.1

$$g(A(t)) \leq v \equiv 1 + \sup_{t \geq 0} b(t) \ \ (t \geq 0).$$

Let us suppose that

$$\rho = -\sup_{t \geq 0} \alpha(A(t)) = \inf_{t \geq 0} Re\ [a(t)/2 - \sqrt{a^2(t)/4 - b(t)}] > 0.$$

Then we can write

$$\zeta_1 = \frac{1}{\rho^2} + \frac{2v}{\rho^3}.$$

So if $\zeta_1 q_0 < 1$, then the considered equation is exponentially stable due to Corollary 3.1.4.

3.2 Proofs of Theorems 3.1.1 and 3.1.5, and Lemma 3.1.7

Proof of Theorem 3.1.1: Equation (1.1) can be rewritten in the form

$$dx/dt - A(\tau)x = [A(t) - A(\tau)]x$$

with an arbitrary fixed $\tau \geq 0$. This equation is equivalent to the following one:

$$x(t) = exp[A(\tau)t]x(0) + \int_0^t exp[A(\tau)(t-t_1)][A(t_1) - A(\tau)]x(t_1)dt_1.$$

According to (1.2),

$$\|x(t)\| \leq p(t, A(\tau))\|x(0)\| + \int_0^t p(t-t_1, A(\tau))q(\tau, t_1)\|x(t_1)\|dt_1.$$

With $\tau = t$ this relation gives

$$\sup_{0\leq t\leq T} \|x(t)\| \leq \chi\|x(0)\| + \sup_{0\leq t\leq T} \|x(t)\|\theta$$

for any positive finite T. By the condition $\theta < 1$ we arrive at the inequality

$$\sup_{0\leq t\leq T} \|x(t)\| \leq \chi\|x(0)\|(1-\theta)^{-1}.$$

Since the right-hand part of the latter inequality does not depend on T, we get

$$\sup_{t\geq 0} \|x(t)\| \leq \chi\|x(0)\|(1-\theta)^{-1}.$$

This is the required inequality for $s = 0$. It can be similarly proved for any $s \leq t$. □

Set

$$\zeta_0 := \int_0^\infty \tilde{p}_0(t)dt.$$

Simple calculations show that

$$\zeta_0 = \sum_{k=0}^{n-1} \frac{v^k}{\sqrt{k!}\rho^{1+k}}.$$

Lemma 3.2.1 *Let the conditions (1.5), (1.8), (1.9) and* $q_0\zeta_1 < 1$ *be fulfilled. Then the evolution operator* $U(t,s)$ *of (1.1) satisfies the estimate*

$$\int_0^t \|U(t,w)\|dw \leq \frac{\zeta_0}{1-q_0\zeta_1} \text{ for all } t \geq 0.$$

Proof: Again rewrite equation (1.1) in the form

$$dx/dt - A(\tau)x = [A(t) - A(\tau)]x$$

with a fixed $\tau \geq 0$. Hence

$$x(t) = exp[A(\tau)t]x(w) + \int_w^t exp[A(\tau)(t-t_1)][A(t_1) - A(\tau)]x(t_1)dt_1$$

$$(0 \leq w \leq t < \infty).$$

Using relation (1.5), we get

$$\|x(t)\| \leq \|e^{A(\tau)(t-w)}x(w)\| + \int_w^t \|e^{A(\tau)(t-t_1)}\| \, q_0|t_1 - \tau|\|x(t_1)\|dt_1.$$

Hence, with the notation $p_0(t) \equiv \tilde{p}_0(t)$, we have

$$\|x(t)\| \leq p_0(t-w)\|x(w)\| + \int_w^t p_0(t-t_1)q_0|t_1 - \tau|\|x(t_1)\|dt_1.$$

When $\tau = t$ this relation implies the inequality

$$y(t,w) \leq p_0(t-w) + q_0 \int_0^t p_0(t-t_1)(t-t_1)y(t_1,w)dt_1$$

with

$$y(t,w) = \frac{\|x(t)\|}{\|x(w)\|}.$$

The integration with respect to w yields

$$\int_0^t y(t,w)dw \leq \zeta_0 + q_0 \int_0^t \int_w^t p_0(t-t_1)(t-t_1)y(t_1,w)dt_1 dw, \tag{2.1}$$

because

$$\int_0^t p_0(t-w)dw \leq \int_0^\infty p_0(s)ds = \zeta_0.$$

Obviously,

$$\int_0^t \int_w^t p_0(t-u)(t-u)y(u,w)du\,dw =$$

$$\int_0^t \int_0^u p_0(t-u)(t-u)y(u,w)dw\,du.$$

Hence,

$$\int_0^t \int_w^t p_0(t-u)(t-u)y(u,w)dudw \leq$$

$$\int_0^t p_0(t-u)(t-u)du \max_{u\le t} \int_0^u y(u,w)dw.$$

Now, due to (2.1) we can write down

$$\int_0^t y(t,w)dw \le \zeta_0 + q_0 \max_{u\le t} \int_0^u y(u,w)dw \int_0^t p_0(t-u)(t-u)du \le$$

$$\zeta_0 + q_0\zeta_1 \max_{u\le t} \int_0^u y(u,w)dw.$$

For any fixed positive T, this implies

$$\max_{0\le t\le T} \int_0^t y(t,w)dw \le \zeta_0 + q_0\zeta_1 \max_{0\le u\le T} \int_0^u y(u,w)dw.$$

Thus, due to the inequality $q_0\zeta_1 < 1$, it can be written

$$\max_{0\le t\le T} \int_0^t y(t,w)dw \le (1-q_0\zeta_1)^{-1}\zeta_0.$$

The right hand part of the latter inequality does not depend on T. Therefore,

$$\max_{t\ge 0} \int_0^t y(t,w)dw \le (1-\zeta_1)^{-1}\zeta_0.$$

This result proves the lemma, since $x(t)$ is an arbitrary solution. □

Proof of Theorem 3.1.5: Under the hypothesis of the theorem we have the inequality $\rho > z(q_0, v)$. Dividing (1.10) by z^{n+1}, and taking into account that $z(q_0, v)$ is the extreme right-hand root of (1.10), we can write down

$$1 > q_0 \sum_{k=0}^{n-1} \frac{(k+1)v^k}{\sqrt{k!}\rho^{k+2}} \equiv q_0\zeta_1.$$

Now, Corollary 3.1.5 provides the stability.

Let us prove the upper solution estimate. Substituting the equality

$$x(t) = x_c(t)exp(-tc)$$

with a real number c in equation (1.1), we get the equation

$$dx_c/dt = (A(t) + Ic)x_c. \tag{2.2}$$

Take $c = \rho - z(q_0, v) - \epsilon$ with an $\epsilon > 0$. Then we can affirm that the matrix

$$A(t) + I(z(q_0, v) + c)$$

is a Hurwitz one. As it is shown in Section 1.5,

$$g(A(t) + Ic) = g(A(t)).$$

Therefore, due to Corollary 3.1.4 equation (2.2) is stable, i.e. there is a constant a such that

$$\|x_c(t)\| \le a\|x_c(0)\| \text{ for all } t \ge 0.$$

This ensures the upper estimate, since $x(t) = x_c(t)\exp(-tc)$, and ϵ is arbitrary.

Now let us prove the lower estimate. Take into account that the Cauchy operator $U(t)$ of equation (1.1) has the following property: its inverse operator $U^{-1}(t)$ is the Cauchy operator of the equation

$$\dot{u} = -A(t)u.$$

But

$$\sup_{t\ge 0}\max_k \; Re\lambda_k(-A(t)) = -\inf_{t\ge 0}\min_k \; Re\lambda_k(A(t)).$$

Due to the proven upper estimate

$$\|U^{-1}(t)h\| \le a\|h\|exp\,[(-\beta_0 + z(q_0, v))t],$$

since $g(-A) = g(A)$. Putting $w = U^{-1}(t)h$, we arrive at the relation

$$\|U(t)w\| \ge a^{-1}\|exp\,[(\beta_0 - z(q_0, v))t]\;\|w\| \; (w \in \mathbf{C}^n).$$

That inequality gives the lower estimate. □

Proof of Lemma 3.1.7: Take into account that by (1.10),

$$q_0 = \frac{z^{n+1}(q_0, v)}{P(z(q_0, v))}.$$

But the function $x^{n+1}P^{-1}(x)$ increases as $x > 0$ increases. Hence by (1.11), $\rho > z(q_0, v)$. This fact yields the stated result. □

3.3 Systems with Differentiable Matrices

Let $A(t) = (a_{jk}(t))_{j,k=1}^n$ be a real differentiable matrix, which is Hurwitzian for each $t \ge 0$ and uniformly bounded on $[0, \infty)$. In this section we establish stability conditions for a linear system in terms of the determinant of the variable matrix.

Denote

$$\psi(t) = \frac{1}{\pi}\int_{-\infty}^{\infty} \|(A(t) - iI\omega)^{-1}\|^3 d\omega \;\; (t \ge 0).$$

Theorem 3.3.1 *Let*

$$\|\dot{A}(t)\|\psi(t) \leq 1 \ (t \geq 0). \tag{3.1}$$

Then system (1.1) is stable.

This theorem and the next one are proved in Section 3.5. Now let

$$\sup_{t \geq t_0} \|\dot{A}(t)\|\psi(t) < 1. \tag{3.2}$$

With an $\epsilon > 0$ substitute the equality $x(t) = u_\epsilon(t)e^{-\epsilon t}$ into (1.1). Then we have the equation

$$\dot{u}_\epsilon(t) = (A(t) + I\epsilon)u_\epsilon(t) \ \ (t \geq 0).$$

Put

$$\psi_\epsilon(t) = \frac{1}{\pi}\int_{-\infty}^{\infty} \|(A(t) - (1-\epsilon)I\omega)^{-1}\|^3 d\omega \ \ (t \geq 0).$$

Due to (3.2) we can take ϵ sufficiently small such that $\|\dot{A}(t)\|\psi_\epsilon(t) \leq 1 \ (t \geq t_0)$. Thanks to Lemma 3.3.1, there is a constant M_ϵ, such that

$$\|u_\epsilon(t)\| \leq M_\epsilon \|x(0)\| \ \ (t \geq t_0).$$

Thus

$$\|x(t)\| \leq M_\epsilon \|x(0)\| e^{-\epsilon t} \ \ (t \geq t_0).$$

This implies

Corollary 3.3.2 *Let condition (3.2) hold. Then system (1.1) is exponentially stable.*

Put

$$\tilde{\psi}(t) := \frac{1}{\pi}\int_{-\infty}^{\infty} \frac{(N^2(A(t)) + n\omega^2)^{3(n-1)/2} d\omega}{|det\ (i\omega I - A(t))|^3 \ \ (n-1)^{3(n-1)/2}}.$$

Recall that $N(A_0)$ is the Frobenius norm of a matrix A_0.

Theorem 3.3.3 *For some $t_0 \geq 0$, let*

$$\|\dot{A}(t)\|\tilde{\psi}(t) \leq 1 \ (t \geq t_0).$$

Then system (1.1) is stable. Moreover, if

$$\sup_{t \geq t_0} \|\dot{A}(t)\|\tilde{\psi}(t) < 1.$$

Then system (1.1) is exponentially stable.

Note that if all the eigenvalues of the matrix are real, then

$$|det\,(i\omega I - A(t))| \geq (\omega^2 + \alpha^2(t))^{n/2} \quad (\omega \in \mathbf{R}),$$

where

$$\alpha(t) := \alpha(A(t)) = \max_{k=1,\dots,n} Re\,\lambda_k(A(t))$$

and $\lambda_k(A(t))$ $(k = 1, ..., n)$ are the eigenvalues of an $n \times n$-matrix $A(t)$ with their multiplicities. Thus,

$$\tilde{\psi}(t) \leq \frac{\gamma_n}{\pi} \int_0^\infty \frac{(N^2(A(t)) + ny^2)^{3(n-1)/2}}{(y^2 + \alpha^2(t))^{3n/2}} dy,$$

where

$$\gamma_n = \frac{1}{(n-1)^{3(n-1)/2}}.$$

Hence,

$$\tilde{\psi}(t) \leq \frac{\gamma_n}{\pi|\alpha(t)|^{3n-1}} \int_0^\infty \frac{(N^2(A(t)) + n\alpha^2(t)s^2)^{3(n-1)/2}}{(s^2+1)^{3n/2}} ds.$$

Take into account that

$$\sum_{k=1}^n |\lambda_k(A(t))|^2 \leq N^2(A(t))$$

and

$$|\alpha(t)| \leq \min_k |\lambda_k(A(t))|.$$

So

$$n\alpha^2(t) \leq N^2(A(t))$$

and

$$\tilde{\psi}(t) \leq \frac{2^{3(n-1)/2}\gamma_n N^{3(n-1)}(A)}{\pi|\alpha(t)|^{3n-1}} \int_0^\infty \frac{1}{(s^2+1)^{3/2}} dy.$$

It can be checked by differentiation that

$$\int_0^t \frac{1}{(s^2+1)^{3/2}} ds = \frac{t}{(t^2+1)^{1/2}}.$$

Thus $\tilde{\psi}(t) \leq \psi_0(t)$, where

$$\psi_0(t) = \frac{2^{3(n-1)/2}\gamma_n N^{3(n-1)}(A(t))}{\pi|\alpha(t)|^{3n-1}}.$$

In the case $n = 2$,

$$\psi_0(t) = \frac{4\sqrt{2}N^3(A(t))}{\pi|\alpha(t)|^5}. \qquad (3.3)$$

Now Theorem 3.3.3 implies

Corollary 3.3.4 *For a $t_0 \geq 0$, let all the eigenvalues $A(t)$ be real, and*

$$\|\dot{A}(t)\|\psi_0(t) \leq 1 \quad (t \geq t_0).$$

Then system (1.1) is stable.

Example 3.3.5 *Let us consider the system*

$$\dot{x}_1 = -3a(t)x_1 + b(t)x_2, \; \dot{x}_2 = a(t)x_2 - b(t)x_1 \tag{3.4}$$

where $a(t), b(t)$ are positive differentiable functions, and

$$3a^2(t) < b^2(t) < 4a^2(t).$$

Then

$$\lambda_{1,2}(A(t)) = -a(t) \pm \sqrt{4a^2(t) - b^2(t)},$$
$$N^2(A(t)) = 10a^2(t) + 2b^2(t)$$

and by (3.3)

$$\psi_0(t) = \frac{16(5a^2(t) + b^2(t))^{3/2}}{\pi|\alpha(t)|^5}.$$

Moreover, simple calculations show that

$$\|\dot{A}(t)\| \leq 3|\dot{a}(t)| + |\dot{b}(t)|$$

Thanks to Corollary 3.3.4, system (3.4) is stable, provided

$$16(3|\dot{a}(t)| + |\dot{b}(t)|)\frac{(5a^2(t) + b^2(t))^{3/2}}{\pi|\alpha(t)|^5} \leq 1 \quad (t \geq 0).$$

3.4 Additional Stability Conditions

In this section as well as in the previous section, $A(t)$ is a real differentiable matrix, which is Hurwitzian for each $t \geq 0$, but the stability conditions are formulated in terms of the individual eigenvalues.

Denote

$$\eta(t) = \int_0^\infty \|e^{A(t)s}\|^2 \, ds.$$

Theorem 3.4.1 *For a $t_0 \geq 0$, let*

$$\|\dot{A}(t)\|\eta^2(t) \leq 1 \; (t \geq t_0). \tag{4.1}$$

Then system (1.1) is stable. Moreover, if

$$\sup_{t \geq t_0} \|\dot{A}(t)\|\eta^2(t) < 1, \tag{4.2}$$

then system (1.1) is exponentially stable.

This theorem is proved in the next section.

Recall that $g(A)$ is introduced in Section 1.5 and $\alpha(t) = \alpha(A(t))$ is defined in the previous section. Put

$$\tilde{\eta}(t) := \sum_{j,k=0}^{n-1} \frac{g^{j+k}(A(t))\,(k+j)!}{2^{j+k+1}|\alpha(t)|^{j+k+1}\,(j!\,k!)^{3/2}}.$$

Due to Lemma 1.9.2 $\eta(t) \le \tilde{\eta}(t)$. Now the previous theorem implies

Corollary 3.4.2 *For a $t_0 \ge 0$, let*

$$\|\dot{A}(t)\|\tilde{\eta}^2(t) \le 1 \ \ (t \ge t_0).$$

Then system (1.1) is stable. Moreover, if

$$\sup_{t \ge t_0} \|\dot{A}(t)\|\tilde{\eta}^2(t) < 1,$$

then system (1.1) is exponentially stable.

In particular, if $n = 2$, then

$$\tilde{\eta}(t) = \frac{1}{2|\alpha(t)|}(1 + \kappa(t) + \frac{\kappa^2(t)}{2}), \tag{4.3}$$

where

$$\kappa(t) := \frac{g(A(t))}{|\alpha(t)|}.$$

If $n = 3$, then

$$\tilde{\eta}(t) = \frac{1}{2|\alpha(t)|}(1 + \kappa + \frac{(\sqrt{2}+1)}{2}\kappa^2 + \frac{3\sqrt{2}\kappa^3}{4} + \frac{2\kappa^4}{3}) \ \ (\kappa = \kappa(t)). \tag{4.4}$$

Example 3.4.3 *Let $A(t) = (a_{jk}(t))_{j,k=1}^2$ be a real differentiable 2×2 matrix.*

Then Example 1.6.1 implies $g(A(t)) \le |a_{12}(t) - a_{21}(t)|$ in the general case; if the eigenvalues of $A(t)$ are nonreal, then

$$g(A(t)) = \sqrt{(a_{11}(t) - a_{22}(t))^2 + (a_{21}(t) + a_{12}(t))^2}. \tag{4.5}$$

For instance, let us consider the system

$$\dot{x}_1 = -3a(t)x_1 + x_2, \ \dot{x}_2 = a(t)x_2 - x_1, \tag{4.6}$$

where $a(t)$ is a positive differentiable function satisfying the inequality

$$a(t) \le 1/2 \ \ (t \ge 0).$$

Then $\|\dot{A}(t)\| = 3|\dot{a}(t)|$, $\alpha(A(t)) = -a(t)$ and due to (4.5) $g(A(t)) = 4a(t)$. Thus (4.3) implies

$$\tilde{\eta}(t) = \frac{13}{2a(t)}.$$

Thanks to Corollary 3.4.2, system (4.6) is stable, provided

$$\frac{3 \cdot 13^2 |\dot{a}(t)|}{4a^2(t)} \leq 1 \ \ (t \geq 0).$$

Example 3.4.4 *Let $A(t) = (a_{jk}(t))_{j,k=1}^3$ be a real differentiable 3×3-matrix.*

According to property (5.3) from Section 1.5,

$$g(A(t)) \leq v(t) : = [(a_{12} - a_{21})^2 + (a_{13} - a_{31})^2 + (a_{23} - a_{32})^2]^{1/2}.$$

For simplicity assume that

$$a_{jj}(t) + \sum_{k=1, k \neq j}^{3} |a_{jk}(t)| \leq -\rho_0(t) \ \ (t \geq 0, j = 1, 2, 3),$$

where $\rho_0(t)$ is a positive scalar function. Then due to the well known result (Marcus and Minc, 1964, Section 3.3.5), $\alpha(A(t)) \leq -\rho_0(t)$.

Thus (4.4) implies

$$\tilde{\eta}(t) \leq \eta_1(t) : = \frac{1}{2\rho_0(t)}(1 + \kappa_1 + \frac{(\sqrt{2}+1)\kappa_1^2}{2} + \frac{3\sqrt{2}\kappa_1^3}{4} + \frac{2\kappa_1^4}{3}),$$

where $\kappa_1 = \kappa_1(t) = \frac{v(t)}{\rho_0(t)}$. Thanks to Corollary 3.4.2, the considered system is stable, provided

$$\eta_1^2(t) \| \dot{A}(t) \| \leq 1 \ \ (t \geq 0).$$

3.5 Proofs of Theorems 3.3.1, 3.3.3, and 3.4.2

Recall the Lyapunov theorem (see Section 1.9): if the eigenvalues of A_0 lie in the interior of the left half-plane, then for any positive definite Hermitian matrix H there exists a positive definite Hermitian matrix W_H, such that

$$W_H A_0 + A_0^* W_H = -2H. \tag{5.1}$$

Moreover,

$$W_H = \frac{1}{\pi} \int_{-\infty}^{\infty} (-iI\omega - A_0^*)^{-1} H (iI\omega - A_0)^{-1} d\omega.$$

Now let $A(t)$ depend on t. Put

$$W(t) = \frac{1}{\pi} \int_{-\infty}^{\infty} (-iI\omega - A^*(t))^{-1} (iI\omega - A(t))^{-1} d\omega. \tag{5.2}$$

Then $W(t)$ is a solution of the equation.

$$W(t)A(t) + A^*(t)W(t) = -2I. \qquad (5.3)$$

Since $A(.)$ is real,

$$(W(t)A(t)h, h) = -(h, h) \;\; (h \in \mathbf{R}^n).$$

Furthermore, multiplying equation (1.1) by $W(t)$ and doing the scalar product, we get

$$(W(t)\dot{x}(t), x(t)) = (W(t)A(t)x(t), x(t)) = -(x(t), x(t)). \qquad (5.4)$$

Since

$$\frac{d}{dt}(W(t)x(t), x(t)) = (W(t)\dot{x}(t), x(t)) + (\dot{W}(t)x(t), x(t)) + (W(t)x(t), \dot{x}(t)) =$$

$$2(W(t)\dot{x}(t), x(t)) + (\dot{W}(t)x(t), x(t)),$$

it can be written

$$\frac{d}{dt}(W(t)x(t), x(t)) = (\dot{W}(t)x(t), x(t)) + 2(W(t)\dot{x}(t), x(t)).$$

Thus due to (5.4)

$$\frac{d}{dt}(W(t)x(t), x(t)) = (\dot{W}(t)x(t), x(t)) - 2(x(t), x(t)).$$

and the inequality

$$\|\dot{W}(t)\| \le 2 \;\; (t \ge t_0) \qquad (5.5)$$

provides the stability, since

$$(W(t)h, h) \ge c_0(h, h) \; (c_0 = const > 0, \; t \ge 0, \; h \in \mathbf{R}^n).$$

Indeed, the latter relation follows from Lemma 1.10.1 and the uniform boundedness of $A(t)$. We thus have proved

Lemma 3.5.1 *Let condition (5.5) hold. Then system (1.1) is stable.*

Proof of Theorem 3.3.1: From (5.3) it follows

$$\dot{W}(t) = \frac{1}{\pi}\int_{-\infty}^{\infty} (-iI\omega - A^*(t))^{-1}[\dot{A}^*(t)(-iI\omega - A^*(t))^{-1} +$$

$$(-iI\omega - A(t))^{-1}\dot{A}(t)](iI\omega - A(t))^{-1} d\omega.$$

Hence,

$$\|\dot{W}(t)\| \le 2\|\dot{A}(t)\|\frac{1}{\pi}\int_{-\infty}^{\infty} \|(iI\omega - A(t))^{-1}\|^3 d\omega = 2\|\dot{A}(t)\|\psi(t).$$

This and (5.5) prove the required result. □

Proof of Theorem 3.3.3: Recall that for any constant $n \times n$-matrix A_0 the inequality

$$\|(I\lambda - A_0)^{-1}\| \leq \frac{(N^2(A_0) - 2Re\ (\overline{\lambda}\ Trace\ (A_0))\ + n|\lambda|^2)^{(n-1)/2}}{|det\ (\lambda I - A_0)|(n-1)^{(n-1)/2}}$$

is true for any regular λ (see Section 1.5). Since $A(t)$ is real and $Im\ Trace\ A(t) = 0$, we have

$$\|(Iiy - A(t))^{-1}\| \leq \frac{(N^2(A(t)) + ny^2)^{(n-1)/2}}{|det\ (iyI - A(t))|\ (n-1)^{(n-1)/2}}\ (y \in \mathbf{R}).$$

Hence, $\psi(t) \leq \tilde{\psi}(t)$. Now Theorem 3.3.1 implies the required result. □

Proof of Theorem 3.4.1: Due to (5.2) we can write

$$W(t) = 2\int_0^\infty e^{A^*(t)s} e^{A(t)s} ds \tag{5.6}$$

(see Section 1.9). So $W(t)$ is a solution of equation (5.3). We again use the stability condition (5.5). To estimate $\|\dot{W}(t)\|$, let us differentiate equation (5.3). Then

$$\dot{W}(t)A(t) + A^*(t)\dot{W}(t) = -H_1(t),$$

where

$$H_1(t) := W(t)\dot{A}(t) + \dot{A}^*(t)W(t).$$

Due to (5.1) and (5.6)

$$\dot{W}(t) = \int_0^\infty e^{A^*(t)s} H_1(t) e^{A(t)s} ds.$$

Moreover, $\|W(t)\| \leq 2\eta(t)$ and $\|H_1(t)\| \leq 2\|\dot{A}(t)\|\eta(t)$. Hence,

$$\|\dot{W}(t)\| \leq \|H_1(t)\| \int_0^\infty \|e^{A(t)s}\|^2\, ds = \|H_1(t)\|\eta(t) \leq 2\|\dot{A}(t)\|\eta^2(t).$$

Now Lemma 3.5.1 proves the stability.

If (4.2) holds, then repeating the arguments of the proof of Corollary 3.3.2 we get the exponential stability. The proof is complete. □

3.6 Matrix Lipschitz Conditions

3.6.1 Stability Criterion

Consider equation (1.1) in the form

$$\frac{dx_j}{dt} = \sum_{k=1}^{n} a_{jk}(t)x_k \ (t \geq 0;\ j = 1, ..., n) \tag{6.1}$$

under the matrix Lipschitz conditions

$$|a_{jk}(t) - a_{jk}(s)| \leq q_{jk}|t - s| \ (t, s \geq 0;\ j, k = 1, ..., n),$$

where q_{jk} are constants. We will write these conditions in the form

$$|A(t) - A(s)| \leq Q|t - s| \ \ (t, s \geq 0), \tag{6.2}$$

where

$$A(t) = (a_{jk}(t))_{j,k=1}^{n} \text{ and } Q = (q_{jk})_{j,k=1}^{n}$$

is a constant matrix. Inequality (6.2) permits us to obtain estimates for each coordinate of a solution.

Throughout the present section again it is assumed that matrix $A(t)$ is uniformly bounded on $[0, \infty)$.

Furthermore, let $m_{ij}(\lambda, t)$ $(\lambda \in \mathbf{C},\ t \geq 0)$ be the cofactor of the element $\delta_{ij}\lambda - a_{ij}(t)$ of the matrix $\lambda I - A(t)$. Recall that $\delta_{ij} = 0,\ j \neq i$ and $\delta_{jj} = 1$. Assume that after division by the coinciding zeros we obtain the equality

$$\frac{m_{ij}(\lambda, t)}{det(I\lambda - A(t))} = \frac{\mu_{ij}(\lambda, t)}{d_{ij}(\lambda, t)},$$

where $d_{ij}(\lambda, t)$ and $\mu_{ij}(\lambda, t)$ are polynomials with respect to λ, and the degree of $d_{ij}(\lambda, t)$ is denoted by $n_{ij}(t)$. Let $co(A(t))$ be the convex hull of all eigenvalues of matrix $A(t)$. Put

$$b_{ji}(k, t) = \frac{1}{k!(n_{ij} - 1 - k)!} \sup_{\lambda \in co(A(t))} |\frac{\partial^{(n_{ij}-k-1)} \mu_{ij}(\lambda, t)}{\partial \lambda^{(n_{ij}-k-1)}}|,$$

$$(n_{ij} \equiv n_{ij}(t);\ k = 0, ..., n - 1).$$

Since $(-m)! = 0$ for a natural $m > 0$, the equalities

$$b_{ij}(k, t) = 0 \text{ for } k \geq n_{ij}(t)$$

are valid if $n_{ij}(t) < n - 1$. Set

$$\bar{b}_{ij}(k) = \sup_{t \geq 0} b_{ij}(k, t)$$

and

$$\overline{B}_k = (\bar{b}_{ij}(k))_{j,i=1}^n.$$

Finally, denote

$$\alpha_0 \equiv \sup_{t \geq 0} \max_{k=1,\ldots,n} Re\lambda_k(A(t)).$$

Theorem 3.6.1 *If $A(t)$ satisfies condition (6.2), then any solution $x(t)$ of equation (1.1) subordinates the inequality*

$$|x(t)| \leq exp[(\alpha_0 + z_0)t]C(t)|x(0)| \ (t \geq 0),$$

where $C(t)$ is a non-negative polynomial matrix, and z_0 is the extreme right-hand (positive) root of the polynomial

$$det(I\lambda^{n+1} - \sum_{k=0}^{n-1} \lambda^{n-k-1}(k+1)!\overline{B}_k Q). \tag{6.3}$$

Below in this section we will derive a simple estimate for z_0. Recall that I is the unit matrix.

Corollary 3.6.2 *Under condition (6.2), let the matrix $A(t) + (z_0 + \epsilon)I$ be a Hurwitz one for an $\epsilon > 0$ and all $t \geq 0$. Then system (6.1) is exponentially stable.*

Example 3.6.3 *Let us consider the equation*

$$y'' + a(t)y' + y = 0, \tag{6.4}$$

where the scalar-valued function $a(t)$ satisfies the conditions

$$|a(t) - a(s)| \leq q|t - s| \text{ and } 0 < a(t) \leq 2 \ (t, s \geq 0).$$

Reducing this equation to the form (6.1) by the substitution $x_1 = y'$ and $x_2 = y$, we find that

$$\alpha_0 = -\frac{1}{2} \inf_{t \geq 0} a(t),$$

and $co(A(t)) = [\lambda_1(A(t)), \lambda_2(A(t))]$ is the segment with

$$\lambda_1(A(t)) = -a(t)/2 + i\sqrt{1 - a^2(t)/4}$$

and

$$\lambda_2(A(t)) = -a(t)/2 - i\sqrt{1 - a^2(t)/4}.$$

We have

$$Q = \begin{pmatrix} q & 0 \\ 0 & 0 \end{pmatrix}$$

and

$$\overline{B}_k Q = \begin{pmatrix} q\bar{b}_{11}(k) & 0 \\ q\bar{b}_{21}(k) & 0 \end{pmatrix}$$

with $k = 0, 1$. Clearly, $m_{11}(\lambda, t) = -\lambda$ and $m_{12}(\lambda, t) = -1$. Hence $\bar{b}_{21}(1) = 1, \bar{b}_{21}(0) = 0$,

$$\bar{b}_{11}(1) = \max_{\lambda \in co(A(t))} |\lambda| = 1,$$

and $\bar{b}_{11}(0) = 1$. Thus

$$\Delta(\lambda) := det\ (\lambda^3 I - \lambda \overline{B}_0 Q - 2\overline{B}_1 Q) =$$

$$det \begin{pmatrix} \lambda^3 - \lambda q - 2q & 0 \\ -q & \lambda^3 \end{pmatrix} = \lambda^3(\lambda^3 - \lambda q - 2q).$$

The extreme right-hand root z_0 of $\Delta(\lambda)$ coincides with the the extreme right-hand root of the polynomial $\lambda^3 - \lambda q - 2q$. The Cardan formula yields

$$z_0 = (q + \sqrt{q^2 - (q/3)^3}\ )^{1/3} + (q - \sqrt{q^2 - (q/3)^3}\ )^{1/3}\ (q \le 27).$$

Cubing this expression and making elementary calculations, we find that

$$z_0^3 = q(2 + (q + \sqrt{q^2 - (q/3)^3})^{1/3} + (q - \sqrt{q^2 - (q/3)^3})^{1/3}).$$

It is simple to check that the function

$$(q + \sqrt{q^2 - (q/3)^3})^{1/3} + (q - \sqrt{q^2 - (q/3)^3})^{1/3}$$

monotonically increases as $q < 27$ increases. Hence for an arbitrary fixed $q \le 27$ we have

$$z_0 \le \sqrt[3]{2q(1 + q^{1/3})}.$$

Theorem 3.6.1 implies the following stability condition of equation (6.4):

$$-a(t)/2 + \sqrt[3]{2q(1 + q^{1/3})} < 0\ \ (t \ge 0,\ q \le 27).$$

3.6.2 Proof of Theorem 3.6.1

Let $T_1, T_2, ..., T_m$ be arbitrary constant $n \times n$-matrices. Recall that a number ω is called a characteristic value of the matrix pencil

$$\Phi(\lambda) = \lambda^m I - \lambda^{m-1} T_1 - ... - T_m \tag{6.5}$$

if there is a non-zero vector h_0 such that $\Phi(\omega) h_0 = 0$. We will use the following

Lemma 3.6.4 *Each characteristic value of the matrix pencil defined by (6.5) is an eigenvalue of the block matrix*

$$\begin{pmatrix} T_1 & T_2 & \dots & T_{m-1} & T_m \\ I & 0 & \dots & 0 & 0 \\ . & . & \dots & . & . \\ 0 & 0 & \dots & I & 0 \end{pmatrix}.$$

For the proof see (Rodman, 1989, Chapter 1). By virtue of the Frobenius theorem for positive matrices (Gantmaher, 1967), we arrive at the following result.

Corollary 3.6.5 *Let all the matrices T_k $(k = 1, ..., m)$ be non-negative. Then the extreme right characteristic value of the matrix pencil (6.5) is positive.*

Proof of Theorem 3.6.1: Fix $s \geq 0$ and write (6.1) as

$$dx/dt - A(s)x = [A(t) - A(s)]x.$$

This equation is equivalent to the following one:

$$x(t) = e^{A(s)t}x(0) + \int_0^t e^{A(s)(t-\tau)}[A(\tau) - A(s)]x(\tau)d\tau.$$

Hence (6.2) implies that

$$|x(t)| \leq |e^{A(s)t}x(0)| + \int_0^t |e^{A(s)(t-\tau)}|Q|\tau - s||x(\tau)|d\tau.$$

Putting $t = s$ and writing t instead of s, we find that

$$|x(t)| \leq |e^{A(t)t}x(0)| + \int_0^t (t-\tau)|e^{A(t)(t-\tau)}|Q|x(\tau)|d\tau.$$

According to Corollary 1.12.2,

$$|e^{A(\tau)t}| \leq e^{\alpha_0 t}\Psi(t) \quad (\tau, t \geq 0),$$

where

$$\Psi(t) = \sum_{k=0}^{n-1} t^k \overline{B}_k.$$

Consequently,

$$|x(t)|e^{-\alpha_0 t} \leq \Psi(t)|x(0)| + \int_0^t \Psi(t-\tau)(t-\tau)Qe^{-\alpha_0\tau}|x(\tau)|d\tau.$$

It follows from Lemma 1.7.1 that

$$|x(t)|e^{-\alpha_0 t} \leq \eta(t), \tag{6.6}$$

where $\eta(t)$ is the solution of the equation

$$\eta(t) = \Psi(t)|x(0)| + \int_0^t \Psi(t-\tau)(t-\tau)Q\eta(\tau)d\tau.$$

To solve the foregoing equation we apply the Laplace transformation. Let λ be the dual variable, $\overline{\eta}(\lambda), \overline{\Psi}(\lambda)$ and $\overline{L}(\lambda)$ be the Laplace transforms of

$\eta(t)$, $\Psi(t)$ and $t\Psi(t)$, respectively. Taking the Laplace transform of the latter equation and using the fact that the transform of a convolution is the product of the functions in the convolution, we obtain

$$\overline{\eta}(\lambda) = \overline{\Psi}(\lambda)|x(0)| + \overline{L}(\lambda)Q\overline{\eta}(\lambda).$$

By the inverse Laplace transform,

$$\eta(t) = \frac{1}{2\pi i}\int_{Re\lambda > z_0} e^{t\lambda}(I - \overline{L}(\lambda)Q)^{-1}\overline{\Psi}(\lambda)|x(0)|d\lambda. \tag{6.7}$$

Obviously,

$$\overline{\Psi}(\lambda) = \sum_{k=0}^{n-1} \frac{k!}{\lambda^{k+1}}\overline{B}_k$$

and

$$\overline{L}(\lambda) = \sum_{k=0}^{n-1} \frac{(k+1)!}{\lambda^{k+2}}\overline{B}_k Q.$$

These relations yield

$$\lambda^{n+1}\overline{\Psi}(\lambda) = \sum_{k=0}^{n-1} k!\lambda^{n-k}\overline{B}_k,$$

and

$$\lambda^{n+1}\overline{L}(\lambda) = \sum_{k=0}^{n-1}(k+1)!\lambda^{n-k-1}\overline{B}_k Q.$$

By (6.7) we conclude that

$$\eta(t) = \frac{1}{2\pi i}\int_{Re\lambda > z_0} e^{t\lambda}T^{-1}(\lambda)\sum_{k=0}^{n-1} k!\lambda^{n-k}\overline{B}_k \, d\lambda \, |x_0| \tag{6.8}$$

with the notation

$$T(\lambda) = I\lambda^{n+1} - \sum_{k=0}^{n-1}(k+1)!\lambda^{n-k-1}\overline{B}_k Q.$$

Take into account that

$$T^{-1}(\lambda) = \frac{T_1(\lambda)}{det\, T(\lambda)},$$

where $T_1(\lambda)$ is a polynomial matrix. By virtue of Corollary 3.6.4, z_0 is a positive number. Now according to the residue theorem, (6.8) implies that

$$\eta(t) \le exp(z_0 t)[C_0 + C_1 t + .. + C_m t^{m-1}]|x(0)|,$$

where C_k $(k = 0, ..., m)$ are constant matrices and m is the multiplicity of the root z_0. The assertion of the theorem now follows from the latter inequality and inequality (6.6). □

3.6.3 Estimates for z_0

Lemma 3.6.6 *Let z_1 be the extreme right-hand (positive) characteristic value of the matrix pencil*

$$\lambda^n I - \sum_{k=0}^{n-1} \lambda^{n-k-1} Q_k,$$

where Q_k are non-negative matrices. Then $z_1 \leq z_2$, where z_2 is the extreme right-hand root of the algebraic equation

$$\lambda^n = \sum_{k=0}^{n-1} \lambda^{n-k-1} \|Q_k\|. \tag{6.9}$$

Moreover, if

$$\sum_{k=0}^{n-1} \|Q_k\| \leq 1, \tag{6.10}$$

then

$$z_1^n \leq \sum_{k=0}^{n-1} \|Q_k\|. \tag{6.11}$$

Proof: Clearly,

$$1 \leq \sum_{k=0}^{n-1} z_1^{-k-1} \|Q_k\|.$$

On the other hand, dividing (6.9) by λ^n and taking into account that z_2 is the root of (6.9), we arrive at the equality

$$1 = \sum_{k=0}^{n-1} z_2^{-k-1} \|Q_k\|.$$

The comparison of the two last relations gives the inequality $z_1 \leq z_2$. Now inequality (6.11) is due to Lemma 1.8.1. □

Corollary 3.6.7 *Let z_0 be the (positive) extreme right-hand root of the polynomial (6.3), and let*

$$\sum_{k=0}^{n-1} (k+1)! \|\overline{B}_k Q\| \leq 1. \tag{6.12}$$

Then $z_0 \leq c(Q)$, where

$$c(Q) = \sqrt[n+1]{\sum_{k=0}^{n-1} (k+1)! \|\overline{B}_k Q\|}.$$

The latter corollary and Corollary 3.6.2 imply

Corollary 3.6.8 *Under (6.2) and (6.12), for an* $\epsilon > 0$ *and all* $t \geq 0$, *let* $A(t) + I(c(Q) + \epsilon)$ *be a Hurwitz matrix. Then system (6.1) is exponentially stable.*

3.7 Lower Solution Estimates

Theorem 3.7.1 *If (6.2) holds and*

$$\beta_0 = \inf_{t \geq 0} \min_{k=1,\ldots,n} Re\lambda_k(A(t)) > -\infty,$$

then any solution $x(t)$ *of system (6.1) is subject to the estimate*

$$|x(t)| \geq exp[(\beta_0 - z_0)t]C^{-1}(t)|x(0)|, \ (t \geq 0),$$

where $C(t)$ *is a non-negative polynomial matrix, and* z_0 *is the extreme right-hand root of the polynomial (6.3).*

Proof: Fox a fixed $T > 0$, put in (6.1) $\tau = T - t$, $x(T - \tau) = y(\tau)$ and $-A(T - \tau) = A_1(\tau)$. We get

$$dy/d\tau = A_1(\tau)y(\tau) \ (0 \leq \tau \leq T).$$

The eigenvalues of $A_1(\tau)$ are $-\lambda_1(A(T - \tau)), ..., -\lambda_n(A(T - \tau))$. Hence,

$$\sup_{0 \leq \tau \leq T} \max_{k=1,\ldots,n} Re\lambda_k(A_1(\tau)) \leq -\beta_0.$$

Moreover, the closed convex hull of the numbers $-\lambda_1(A(T-\tau)), ..., -\lambda_n(A(T-\tau))$ is the mirror image of $co(A(t))$ in the imaginary axis.

Furhermore, the cofactor of the element $-a_{ij}(T - \tau) - \delta_{ij}\lambda$ of $A_1(\tau) - \lambda I$ equals $-m_{ij}(-\lambda, T - \tau)$, i.e.

$$\frac{1}{k!(n_{ij} - k - 1)!} \sup_{\lambda \in co(-A(T-\tau))} \left|\frac{\partial^{(n_{ij}-k-1)}\mu_{ij}(\lambda, T - t)}{\partial\lambda^{(n_{ij}-k-1)}}\right| \leq \bar{b}_{ji}(k)$$

$$(n_{ij} = n_{ij}(t); \ 0 \leq \tau \leq T; \ i, j = 1, ..., n)$$

(see Subsection 3.6.1). Therefore, Theorem 3.6.1 implies that

$$|y(\tau)| \leq exp[(-\beta_0 + z_0)\tau]C(\tau)|y(0)| \ (0 \leq \tau \leq T).$$

Hence, we obtain the desired inequality. □

4. Dissipative and Piecewise Constant Systems

4.1 The Lozinskii Inequality

Let $\|.\|_{C^n}$ be an arbitrary norm in $\mathbf{C}^n$. There are an infinite number of norms on $\mathbf{C}^n$. However, the following are most commonly used in practice:

$$\|h\|_p = [\sum_{k=1}^{n} |h_k|^p]^{1/p} \quad (1 \le p < \infty)$$

and $\|h\|_\infty = \max_{k=1,\dots,n} |h_k| \;\; (h = (h_k) \in \mathbf{C}^n)$. So $\|.\|_2 = \|.\|$ is the Euclidean norm. Again consider the system

$$\dot{x} = A(t)x \quad (t > 0) \tag{1.1}$$

with a piecewise continuous matrix $A(t)$.

Theorem 4.1.1 *Let there be a piecewise continuous function $\phi(t),\ t \ge 0$, such that for all sufficiently small $\delta > 0$ the condition*

$$\|I + A(t)\delta\|_{C^n} \le 1 + \phi(t)\delta \quad (t \ge 0) \tag{1.2}$$

is fulfilled. Then any solution $x(t)$ of (1.1) satisfies the inequality

$$\|x(t)\|_{C^n} \le \|x(s)\|_{C^n}\, exp[\int_s^t \phi(s_1)ds_1] \quad (t \ge s \ge 0). \tag{1.3}$$

Proof: Condition (1.2) implies

$$\| \overleftarrow{\prod_{1\le k\le m}} (1 + A(t_k^{(m)})\delta_k)\|_{C^n} \le \prod_{k=0}^{m}(1 + \phi(t_k^{(m)})\delta_k).$$

Here $s = t_1^{(m)} < t_2^{(m)} < \dots < t_m^{(m)} = t,\ \delta_k = t_{k+1}^{(m)} - t_k^{(m)}\ (k = 1, \dots, m-1)$. Recall that the arrow over the symbol of the product means that the indexes of the co-factors increase from right to left.

Clearly, $1 + \phi(t_k^{(m)})\delta_k = e^{\phi(t_k^{(m)})\delta_k} + o(\delta_k)\ \ (\delta_k \downarrow 0)$. That is,

$$\| \overleftarrow{\prod_{1\le k\le m}} (I + A(t_k^{(m)})\delta_k)\|_{C^n} \le \exp[\sum_{k=1}^{m} \phi(t_k^{(m)})\delta_k + 0(\delta)] \quad (\delta \downarrow 0),$$

where $\delta = \max_k \delta_k$. The passage to the limit as $m \to \infty$ and Lemma 2.3.1 give us the required estimate. □

Corollary 4.1.2 *(Lozinskii). Let the relation*

$$\overline{\lim}_{\delta \downarrow 0} \frac{\|I + A(t)\delta\|_{C^n} - 1}{\delta} = \phi(t) \ (t \geq 0)$$

hold. Then for any solution $x(t)$ of (1.1) the inequality (1.3) is true.

Indeed, $\|I + A(t)\delta\|_{C^n} = 1 + \phi(t)\delta + o(\delta)$ $(\delta \downarrow 0)$. Now, Theorem 4.1.1 implies inequality (1.3).

We will say that A is a dissipative matrix if $\|I + A\delta\|_{C^n} \leq 1$ for all sufficiently small $\delta > 0$. System (1.1) is said to be dissipative if $A(t)$ is a dissipative matrix for all $t \geq 0$.

Corollary 4.1.3 *Let (1.1) be a dissipative system. Then it is stable.*

Theorem 4.1.4 *Let there be a piecewise continuous function $\psi(t)$, such that for all sufficiently small $\delta > 0$ and $h \in \mathbf{C}^n$, the relation*

$$\frac{\|(I + A(t)\delta)h\|_{C^n}}{\|h\|_{C^n}} \geq 1 + \psi(t)\delta \ \ (t \geq 0)$$

holds. Then any solution $x(t)$ of (1.1) is subject to the inequality

$$\|x(t)\|_{C^n} \geq \|x(s)\|_{C^n} exp[\int_s^t \psi(s_1)ds_1] \ (t, s \geq 0).$$

The proof of this theorem is analogous to the proof of Theorem 4.1.1.

Let $A_R = (A + A^*)/2$ be the real Hermitian component of A; $\beta(A_R)$ and $\alpha(A_R)$ be the smallest and largest eigenvalues of A_R, respectively.

Lemma 4.1.5 *Let $\|.\|$ be the Euclidean norm. Then for any constant $n \times n$-matrix A and any $h \in \mathbf{C}^n$,*

$$1 + \beta(A_R)\delta + o(\delta) \leq \frac{\|(I + A\delta)h\|}{\|h\|} \leq 1 + \alpha(A_R)\delta + o(\delta) \ (\delta \downarrow 0).$$

Proof: It is obvious that

$$\|(I + A\delta)h\|^2 = \|h\|^2 + 2(A_R h, h)\delta + (Ah, Ah)\delta^2 \leq 1 + 2\alpha(A_R)\delta + o(\delta)$$

when $\|h\| = 1$. Here $(.,.)$ is he scalar product. But

$$\sqrt{1 + 2\alpha(A_R)\delta} = 1 + \alpha(A_R)\delta + o(\delta).$$

This clearly forces the upper estimate. The proof of the lower estimate is left to the reader. □

Thus, if we take $\|.\|_{C^n} = \|.\|$-the Euclidean norm, then A is dissipative, if $A + A^*$ is negative definite. Theorems 4.1.1 and 4.1.4, and Lemma 4.1.5 imply

Corollary 4.1.6 *(Wazewski). For any solution $x(t)$ of equation (1.1) and all $t \geq s \geq 0$, the estimates*

$$exp[\int_s^t \beta(A_R(s_1))ds_1] \leq \frac{\|x(t)\|}{\|x(s)\|} \leq exp[\int_s^t \alpha(A_R(s_1))ds_1]$$

are true.

Let E be a positive definite Hermitian matrix. Introduce the scalar product $(.,.)_E$ by $(h,g)_E = (Eh,g)$ $(h,g \in \mathbf{C}^n)$. Put $\|h\|_E = \sqrt{(Eh,h)}$. Theorem 4.1.6 yields

Corollary 4.1.7 *Let $EA^*(t) + A(t)E \leq 0$ $(t \geq 0)$. Then equation (1.1) is stable.*

4.2 Linear Systems with Majorants and Minorants

Let $A(t)$ $(t \geq 0)$ be a variable matrix with entries $a_{jk}(t)$ $(j,k = 1,...,n)$. Let there be a real piecewise continuous matrix $M(t) = (m_{jk}(t))_{j,k=1}^n$ with non-negative off diagonal entries such that

$$|I + A(t)\delta| \leq I + M(t)\delta \ (t \geq 0) \tag{2.1}$$

for all sufficiently small positive δ. Inequality (2.1) means that $|a_{jk}(t)| \leq m_{jk}(t)$ for $j \neq k$ and

$$|1 + \delta\, a_{kk}(t)| \leq 1 + \delta\, m_{kk}(t) \ (t \geq 0;\ j,k = 1,...,n).$$

Lemma 4.2.1 *Let condition (2.1) hold. Then any solution $x(t)$ of system (1.1) satisfies the inequality $|x(t)| \leq z(t)$ $(t \geq 0)$, where $z(t)$ is a solution of the equation*

$$\dot{z} = M(t)z \tag{2.2}$$

with the initial condition $z(0) = |x(0)|$.

Proof: Due to condition (2.1) it can be written

$$|\prod_{1\leq k\leq m}^{\leftarrow} (I + A(t_k^{(m)})\delta_k)u_0| \leq \prod_{1\leq k\leq m}^{\leftarrow} (I + M(t_k^{(m)})\delta_k)|u_0|.$$

But according to Lemma 2.3.1, one can assert that the sequences at the left-hand part and at the right-hand part of this inequality tend to the solutions $x(t)$ and $z(t)$ of equations (2.2) and (2.3), respectively. This finishes the proof. □

Corollary 4.2.2 *Under condition (2.1) let $M(t) \equiv M$ be a constant Hurwitz matrix. Then (1.1) is an exponentially stable system.*

Now we will obtain a lower solution estimate. Suppose that that there is a real matrix $P(t)$ with non-negative off diagonal entries such that

$$|I - A(t)\delta| \le I + P(t)\delta \ (t \ge 0) \tag{2.3}$$

for all sufficiently small positive δ.

Furthermore, let $U(t)$ be the Cauchy operator of equation (1.1). Take into account the operator $U^{-1}(t)$ is the Cauchy operator of the equation

$$\dot{x} = -A(t)x. \tag{2.4}$$

By the previous lemma $|U^{-1}(t)| \le Z(t)$, where $Z(t)$ is the Cauchy operator of the equation $\dot{z} = P(t)z$. Now we easily obtain

$$|U(t)h| \ge Z^{-1}(t)|h| \ (h \in \mathbf{C}^n, t \ge 0).$$

But $Z^{-1}(t)$ is the Cauchy operator of the equation

$$\dot{z} = -P(t)z. \tag{2.5}$$

We thus have derived

Lemma 4.2.3 *Under condition (2.3) any solution $x(t)$ of equation (1.1) satisfies the inequality $|x(t)| \ge z_-(t)$ $(t \ge 0)$, where $z_-(t)$ is the solution of equation (2.5) with the initial condition $z_-(0) = |x(0)|$.*

Let (2.3) hold with a constant matrix $P(t) \equiv P$. Then any solution $x(t)$ of (1.1) is subject to the inequality $|x(t)| \ge e^{-Pt}|x(0)|$ $(t \ge 0)$.

4.3 Systems with Piecewise Constant Matrices

Let an $n \times n$-matrix $A(t)$ be piecewise constant:

$$A(t) = A_k \ (t_k \le t < t_{k+1};\ k = 0, 1, 2, ...) \tag{3.1}$$

with constant matrices A_k $(k = 0, 1, 2, ...)$, and $0 = t_0 < t_1 < t_2 ....$ Let $U(t, s)$ be the evolution operator of (1.1). Since $U(t, s)U(s, \tau) = U(t, \tau)$ $(t, s, \tau \ge 0)$, we have

$$U(t_{k+1}, t_j) = U(t_{k+1}, t_k) ...\ U(t_{j+1}, t_j) = U\ (t_{k+1}, t_j) =$$
$$e^{\delta_k A_k}\, e^{\delta_{k-1} A_{k-1}} ...\ e^{\delta_j A_j} \ (k > j \ge 0)$$

with $\delta_k = t_{k+1} - t_k$. Thanks to Corollary 1.5.3 this relation implies

$$\|U\ (t_{k+1}, t_j)\| \le \|e^{\delta_k A_k}\| \|\, e^{\delta_{k-1} A_{k-1}}\| ...\ \|e^{\delta_j A_j}\| \le$$
$$e^{\delta_k \alpha(A_k)}\gamma(A_k, \delta_k)e^{\delta_{k-1}\alpha(A_{k-1})}\gamma(A_{k-1}, \delta_{k-1})...e^{\delta_j \alpha(A_j)}\gamma(A_j, \delta_j)$$

with the notation

$$\gamma(A,t) = \sum_{k=0}^{n-1} t^k \frac{g^k(A)}{(k!)^{3/2}}. \tag{3.2}$$

Consequently,

$$\|U\ (t_{k+1}, t_j)\| \leq exp\,[\sum_{m=j}^{k} \delta_m \alpha(A_m)]\ \prod_{m=j}^{k} \gamma(A_m, \delta_m). \tag{3.3}$$

We thus have derived

Theorem 4.3.1 *Let the conditions (3.1) and*

$$\sup_{k=1,2,\ldots} \alpha(A_k) + \delta_k^{-1}\, ln\, \gamma(A_k, \delta_k) < 0$$

hold. Then system (1.1) is asymptotically stable.

4.4 Perturbations of Systems with Piecewise Constant Matrices

Let an $n \times n$-matrix $A(t)$ be piecewise continuous. With notation (3.2), let

$$\tilde{\rho}_0 := \sup_{0 \leq \tau \leq 1,\ k=0,1,\ldots,} \alpha(A(k)) + \frac{ln\, \gamma(A(k), \tau)}{\tau} < 0. \tag{4.1}$$

Theorem 4.4.1 *Let the conditions (4.1) and*

$$\sup_{0 \leq t \leq 1;\ k=1,2,\ldots} \|A(t+k) - A(k)\| < \frac{1}{\tilde{\rho}_0}$$

hold. Then system (1.1) is stable.

Proof: Set $A_0(t) = A(k)$ $(k \leq t < k+1;\ k = 0,1,2,...)$. Let $U_0(t,s)$ be the evolution operator of the equation $\dot{v} = A_0(t)v$. Then

$$U_0(t,\tau) = e^{(t-k)A(k)}\, e^{A(k-1)} \ldots e^{A(j)(j-\tau)}\ (k \leq t < k+1;\ j-1 \leq \tau < j;\ k > j). \tag{4.2}$$

Corollary 1.5.3 gives us the inequality

$$\|U_0\ (t,s)\| ds \leq e^{(t-s)\alpha(A(k))} \gamma(A(k), (t-s))\ \ (k \leq s \leq t < k+1).$$

Then under (4.1)

$$\|U_0\ (t,s)\| ds \leq e^{-(t-s)\tilde{\rho}_0}\ \ (k \leq s \leq t < k+1).$$

But according to (4.2), this inequality is true for all $t \geq s \geq 0$. Thus

$$\int_0^t \|U_0(t,s)\|ds \le \int_0^t e^{-(t-s)\tilde{\rho}_0} \le$$

$$\int_0^\infty e^{-s\tilde{\rho}_0} ds = \frac{1}{\rho_0}.$$

Now Lemma 2.2.6 implies the required result. □

4.5 General Second Order Vector Systems

Let us consider the system

$$\ddot{x} + 2A(t)\dot{x} + B(t)x = 0 \quad (t > 0), \tag{5.1}$$

with real piecewise continuous $n \times n$-matrices $A(t)$ and $B(t)$. As above

$$A_R = (A + A^*)/2, A_I = (A - A^*)/2i$$

denote *the real and imaginary components* of a matrix A, respectively. Let A_0 be a symmetric matrix. Then we will write $A_0 \ge 0$ ($A_0 > 0$) if it is positive definite (strongly positive definite). For two symmetric matrices A_0, B_0 we write $A_0 \ge B_0$ if $A_0 - B_0 \ge 0$, and $A_0 > B_0$ if $A_0 - B_0 > 0$.

Take the initial conditions

$$x(t) = x_0, \dot{x}(0) = x_1 \quad (x_0, x_1 \in \mathbf{R}^n) \tag{5.2}$$

Assume that there is a constant $m_A > 0$, such that

$$A_R(t) \ge m_A I \text{ and } 0 \le B_R(t) \le m_A(2A_R(t) - m_A I) \quad (t \ge 0), \tag{5.3}$$

where I is the unit matrix, and put

$$T(t) := 2m_A A(t) - B(t). \tag{5.4}$$

Theorem 4.5.1 *Under condition (5.3), let*

$$p_0 := \sup_{t \ge 0} \|T_R(t)\| + \|T_I(t)\| < 2m_A^2. \tag{5.5}$$

Then equation (5.1) is exponentially stable. Moreover, any solution $x(t)$ of problem (5.1), (5.2) satisfies the estimate

$$\|x(t)\| \le c_2\, e^{\Lambda t}(\|x_0\| + \|x_1\|) \quad (t \ge 0) \tag{5.6}$$

where $\Lambda := -m_A + \sqrt{p_0 - m_A^2} < 0$ and the constant c_2 does not depend on the initial vectors.

This theorem is proved in the next section.

Corollary 4.5.2 *Let matrices $A(t)$ and $B(t)$ be symmetric, and there be a constant $m_A > 0$, such that*

$$A(t) \geq m_A I \text{ and } m_A^2 I \leq 2m_A A(t) - B(t) < 2m_A^2 I \quad (t \geq 0).$$

Then equation (5.1) is exponentially stable. Moreover, estimate (5.6) is valid with

$$p_0 = \sup_{t \geq 0} \|2m_A A(t) - B(t)\| < 2m_A^2.$$

Example 4.5.3 *Consider the equation*

$$\ddot{x} + 2a(t)\dot{x} + b(t)x = 0 \tag{5.7}$$

with real positive piecewise continuous scalar functions $a(t)$ and $b(t)$. Let

$$m_a \equiv inf_{t \geq 0} a(t) > 0.$$

Then due to Corollary 4.5.2, equation (5.7) is exponentially stable, provided

$$m_a^2 \leq 2m_a a(t) - b(t) \leq \tilde{p}_0 \ (t \geq 0)$$

with $\tilde{p} < 2m_a^2$. In particular, take

$$a(t) = 2 + \frac{1}{2} sin^2\ (t), b(t) = 3 + \frac{1}{2} cos\ (t).$$

Then $m_a = 2$, and

$$m_a^2 = 4 \leq 2m_a a(t) - b(t) = 8 + 2sin^2\ t - 3 - \frac{1}{2} cos\ t < 2m_a^2 = 8.$$

So equation (5.7) is exponentially stable.

Example 4.5.4 *Let $A(t) = (a_{jk})_{j,k=1}^n$ $(a_{jk} \equiv a_{jk}(t))$ be a real symmetric matrix, having the properties*

$$a_{jj}(t) \geq 9/8, \quad \sum_{k=1, k \neq j}^{n} |a_{jk}| \leq 1/8 \ (t \geq 0, j = 1, ..., n). \tag{5.8}$$

Then

$$\inf_j a_{jj}(t) - \sum_{k=1, k \neq j}^{n} |a_{jk}| \geq 1 \ (t \geq 0).$$

Due to the well-known result from Section III.2.2 of the book (Marcus and Minc, 1964) (see also the Appendix A, Section A1) we conclude that $A(t) \geq I$. So $m_A = 1$. In addition, take $B(t) = diag\ [b_j(t)]_j^n$, such that

$$0 \leq 2a_{jj} - 7/4 < b_j \leq 2a_{jj} - 5/4 \ (b_j = b_j(t);\ j = 1, 2, ...). \tag{5.9}$$

Then

$$2(a_{jj} - \sum_{k=1,k\neq j}^{n} |a_{jk}|) - b_j \geq$$

$$2a_{jj} - b_j - 1/4 \geq 5/4 - 1/4 = 1.$$

Again use the well-known result from Section III.2.2 of the book (Marcus and Minc, 1964) (see also the Appendix A, Section A1). It gives us the inequality $T(t) \geq I$. In addition,

$$2(a_{jj} + \sum_{k=1,k\neq j}^{n} |a_{jk}|) - b_j$$

$$\leq 2a_{jj} - b_j + 1/4 < 7/4 + 1/4 = 2.$$

Thus $T(t) < 2I$. So under (5.8) and (5.9) system (5.1) is exponentially stable due to Corollary 4.5.2.

4.6 Proof of Theorem 4.5.1

Put in (5.1)

$$x(t) = e^{-mt} y(t) \quad (m \equiv m_A). \tag{6.1}$$

Then we have the equation

$$\ddot{y} - 2(m - A(t))\dot{y} - C(t)y = 0$$

with

$$C(t) = 2A(t)m - Im^2 - B(t) = T(t) - m^2 I.$$

Reduce this equation to the system

$$\dot{y}_1 = 2(m - A(t))y_1 + C(t)y_2, \; \dot{y}_2 = y_1.$$

Doing the scalar product, we get

$$\frac{d}{dt}(y_1, y_1) = 4((m - A(t))y_1, y_1) + 2(C(t)y_2, y_1),$$

$$\frac{d}{dt}(y_2, y_2) = 2(y_2, y_1).$$

Since the matrices are real, we can write out

$$((mI - A(t))y_1, y_1) = ((mI - A_R(t))y_1, y_1) \leq 0.$$

So

$$d(y_1, y_1)/dt \leq 2\|C(t)\|\|y_2\|\|y_1\|.$$

Thus

$$d\|y_1(t)\|/dt \le \|C(t)\|\|y_2(t)\|, \; d\|y_2(t)\|/dt = \|y_1(t)\|. \tag{6.2}$$

Furthermore,

$$\|C\| \le \|C_R\| + \|C_I\| = \|T_R - m^2 I\| + \|T_I\| \;\; (C = C(t), T = T(t)).$$

Under the hypothesis of the present theorem C_R is positive. So

$$\|T_R - m^2 I\| = \sup_{v \in R^n, \|v\|=1} (T_R v, v) - m^2 = \|T_R\| - m^2.$$

Thus

$$\|C\| \le \|C_R\| + \|C_I\| = \|T_R - m^2 I\| + \|T_I\| =$$
$$\|T_R\| + \|T_I\| - m_A^2 \le p_0 - m_A^2$$
$$(t \ge 0; \; C = C(t), T = T(t)).$$

Put

$$b_0 := \sqrt{p_0 - m^2}.$$

Then (6.2) implies

$$d\|y_1(t)\|/dt \le b_0^2 \|y_2(t)\|, \; d\|y_2(t)\|/dt \le \|y_1(t)\|.$$

Hence,

$$\|y_k(t)\| \le z_k(t) \;\; (k = 1, 2; \; t \ge 0), \tag{6.3}$$

where $(z_1(t), z_2(t))$ is a solution of the coupled system

$$\dot{z}_1 = b_0^2 z_2, \; \dot{z}_2 = z_1 \;\; (z_1(0) = \|x_1\|, z_2(0) = \|x_0\|).$$

Simple calculations show that

$$|z_k(t)| \le const \; (|z_1| + |z_2|) e^{b_0 t} \; (k = 1, 2)$$

Hence (6.1) and (6.3) yield the required result. □

4.7 Second Order Vector Systems with Differentiable Matrices

Consider in $\mathbf{R}^n$, the system

$$\ddot{u} + B(t)\dot{u} + C^2(t)u = 0 \; (t > 0), \tag{7.1}$$

where $B(t)$ and $C(t)$ are real $n \times n$-matrices.

Theorem 4.7.1 *Let $B(t)$ be a piecewise continuous matrix, and $C(t)$ be a differentiable positive definite one. If, in addition,*

$$K(t) \equiv -C^{-1}(t)(\dot{C}(t) + B(t)C(t)) \tag{7.2}$$

is a dissipative matrix in the Euclidean norm for all $t \geq 0$, then system (7.1) is stable. Moreover, any solution u of (7.1) satisfies the inequality

$$\|u(t)\|^2 + \|C^{-1}(t)\dot{u}(t)\|^2 \leq \|u(0)\|^2 + \|C^{-1}(0)\dot{u}(0)\|^2 \ (t \geq 0). \tag{7.3}$$

Proof: Clearly, system (7.1) is equivalent to the following one

$$\dot{x} = -Bx - C^2 y, \ \dot{y} = x,$$

where $B = B(t), C = C(t)$. Put $x(t) = Cz(t)$. Then the latter system takes the form $C\dot{z} + \dot{C}z = -BCz - C^2 y, \ \dot{y} = Cz$. Hence

$$\dot{z} = K(t)z - Cy, \ \dot{y} = Cz.$$

Let $A(t)$ be the matrix of the obtained system. Then

$$A(t) + A^*(t) = \begin{pmatrix} K(t) + K^*(t) & 0 \\ 0 & 0 \end{pmatrix}.$$

Clearly, if $K(t)$ is dissipative, then $A(t)$ is dissipative, and the required result is due to Lemma 4.1.5. □

Example 4.7.2 *Consider the equation*

$$\ddot{u} + b(t)\dot{u} + c^2(t)u = 0 \tag{7.4}$$

with a positive piecewise continuous scalar-valued function $b(t)$ and a positive differentiable one $c(t)$. By the previous theorem, the inequality

$$b(t) + \dot{c}(t)c^{-1}(t) \geq 0 \ (t \geq 0)$$

ensures the stability of equation (7.4). Moreover, the inequality

$$u^2(t) + (c^{-1}(t)\dot{u}(t))^2 \leq u^2(0) + (c^{-1}(0)\dot{u}(0))^2 \ (t \geq 0)$$

holds for any solution $u(t)$ of equation (7.4).

5. Nonlinear Systems with Autonomous Linear Parts

5.1 Statement of the Main Result

Denote

$$\Omega(r) = \{h \in \mathbf{C}^n : \|h\| \leq r\}$$

and consider the equation

$$\dot{x} = Ax + F(x,t) \ \ (t \geq 0, x = x(t)), \tag{1.1}$$

where A is a constant Hurwitz $n \times n$-matrix and F maps $\Omega(r) \times [0, \infty)$ into $\mathbf{C}^n$ with the property

$$\|F(h,t)\| \leq \nu \|h\| \text{ for all } h \in \Omega(r) \text{ and } t \geq 0. \tag{1.2}$$

Here $\nu = \nu(r) \equiv const > 0$. Put

$$\tilde{\Gamma} := \int_0^\infty \|e^{At}\| dt \text{ and } \tilde{\chi} := \max_{t \geq 0} \|e^{At}\|.$$

Lemma 5.1.1 *Under condition (1.2), let the inequality*

$$\nu \tilde{\Gamma} < 1$$

be valid. Then the zero solution to equation (1.1) is asymptotically stable. Moreover, any vector x_0 satisfying the condition

$$\frac{\tilde{\chi} \|x_0\|}{1 - \nu \tilde{\Gamma}} < r,$$

belongs to a region of attraction of the zero solution and a solution $x(t)$ of (1.1) with

$$x(0) = x_0 \tag{1.3}$$

subordinates the estimate

$$\|x(t)\| \leq \frac{\tilde{\chi} \|x_0\|}{1 - \nu \tilde{\Gamma}} \ \ (t \geq 0).$$

This result immediately follows from Theorem 7.1.1 proved below. Denote

$$\Gamma(A) = \sum_{j=0}^{n-1} \frac{g^j(A)}{|\alpha(A)|^{j+1}\sqrt{j!}}$$

and

$$\chi(A) = \max_{t \geq 0} exp[\alpha(A)t] \sum_{j=0}^{n-1} \frac{g^j(A)t^j}{\sqrt{j!}}.$$

Recall that $g(A)$ is defined in Section 1.5. Corollary 1.5.3 yields $\tilde{\chi} \leq \chi(A)$; $\tilde{\Gamma} \leq \Gamma(A)$.

Now Lemma 5.1.1 implies

Corollary 5.1.2 *Under condition (1.2), let the inequality*

$$\nu\Gamma(A) < 1 \tag{1.4}$$

hold. Then the zero solution to equation (1.1) is exponentially stable. Moreover, any vector x_0 satisfying the condition

$$\frac{\chi(A)\|x_0\|}{1 - \nu\Gamma(A)} < r, \tag{1.5}$$

belongs to a region of attraction of the zero solution. Besides, a solution $x(t)$ of problem (1.1), (1.3) subordinates the estimate

$$\|x(t)\| \leq \frac{\chi(A)\|x_0\|}{1 - \nu\Gamma(A)} \quad (t \geq 0). \tag{1.6}$$

Introduce the algebraic equation

$$z^n = \nu \sum_{j=0}^{n-1} \frac{g^j(A)}{\sqrt{j!}} z^{n-j-1} \tag{1.7}$$

and denote by $z(\nu, A)$ the extreme right-hand (unique positive and simple) zero of this equation.

Theorem 5.1.3 *Under condition (1.2), let the matrix $A + z(\nu, A)I$ be a Hurwitz one. Then the zero solution to equation (1.1) is asymptotically stable. Moreover, inequality (1.4) holds and any vector x_0 satisfying condition (1.5) belongs to a region of attraction of the zero solution, and estimate (1.6) is valid.*

The proof of this theorem is presented in the next section.

Put

$$p_0 = \nu \sum_{j=0}^{n-1} \frac{g^j(A)}{\sqrt{j!}}.$$

Due to Lemma 1.8.1, $z(\nu, A) \leq \delta(\nu, A)$, where

$$\delta(\nu, A) = \begin{cases} p_0^{1/n} & \text{if } p_0 \leq 1 \\ p_0 & \text{if } p_0 > 1 \end{cases} .$$

Now Theorem 5.1.3 implies

Corollary 5.1.4 *Under condition (1.2), let the matrix $A + \delta(\nu, A)I$ be a Hurwitz one. Then the zero solution to equation (1.1) is asymptotically stable. Moreover, inequality (1.4) holds and any vector x_0 satisfying condition (1.5) belongs to a region of attraction of the zero solution, and estimate (1.6) is valid.*

We will say that (1.1) *is a quasilinear equation* if

$$\lim_{h \to 0} \frac{\|F(h,t)\|}{\|h\|} = 0$$

uniformly in t. Recall the following famous result

Theorem 5.1.5 *(Lyapunov) Let (1.1) be a quasilinear equation. Then if A is a Hurwitz matrix, the zero solution of (1.1) is asymptotically stable. Conversely, if A has an eigenvalue in the interior of the right half-plane, then the zero solution of (1.1) is unstable.*

For the proof see (Daleckii and Krein, 1974, Section 7.2).

5.2 Proof of Theorem 5.1.3

We need the following

Lemma 5.2.1 *Let $z(\nu, A)$ be the extreme right-hand zero of the algebraic equation (1.7). Then the condition*

$$\alpha(A) + z(\nu, A) < 0$$

implies inequality (1.4).

Proof: The hypothesis of the lemma entails the relation

$$|\alpha(A)| > z(\nu, A). \tag{2.1}$$

Dividing (1.7) by z^{n+1}, and taking into account that $z(\nu, A)$ is its root, we arrive at the equality

$$1 = \nu \sum_{j=0}^{n-1} \frac{g^j(A)}{z^{j+1}(\nu, A)\sqrt{j!}}.$$

But according to (2.1)

$$\nu \Gamma(A) = \nu \sum_{j=0}^{n-1} \frac{g^j(A)}{|\alpha(A)|^{j+1}\sqrt{j!}} <$$

$$\nu \sum_{j=0}^{n-1} \frac{g^j(A)}{z^{j+1}(\nu, A)\sqrt{j!}} = 1.$$

This proves the stated result. □

The assertion of Theorem 5.1.3 immediately follows from Corollary 5.1.2 and Lemma 5.2.1.

5.3 Stability Conditions in Terms of Determinants

5.3.1 Statement of the Result

Theorem 5.3.1 *Let matrix A be real and Hurwitzian and the conditions (1.2), and*

$$b(A) := \sup_{|y| \leq 2\|A\|} \|(A - Iiy)^{-1}\| < \frac{1}{\nu} \tag{3.1}$$

hold. Then the zero solution to equation (1.1) is asymptotically stable. Moreover, with the notations

$$\psi(A) := [\frac{1}{2\pi} \int_{-\infty}^{\infty} \|(A - Iiy)^{-1}\|^2 dy\,]^{1/2}$$

and

$$\eta(A) := \frac{\sqrt{2(\|A\| + \nu)}\psi(A)}{1 - \nu b(A)}$$

any vector $x_0 \in \mathbf{C}^n$, satisfying the inequality

$$\|x_0\|\eta(A) < r \tag{3.2}$$

belongs to the region of attraction of the zero solution to equation (1.1). Besides a solution x of problem (1.1), (1.3) satisfies the inequalities

$$\sup_{t \geq 0} \|x(t)\| \leq \|x_0\|\eta(A) \tag{3.3}$$

and

$$\|x\|_{L^2} \equiv [\int_0^{\infty} \|x(t)\|^2 dt]^{1/2} \leq \frac{\|x_0\|\psi(A)}{1 - \nu b(A)}. \tag{3.4}$$

The proof of this theorem is given below in this section. Put

$$\phi(z) := \frac{N^{n-1}(A - z\,I)}{(n-1)^{(n-1)/2}|det\,(A - z\,I)|} \quad (z \in \mathbf{C}).$$

Thanks to Lemma 1.5.7, the relation

$$\|(A - zI)^{-1}\| \le \phi(z)$$

holds, provided the matrix $A - z\,I$ is invertible. Recall that $N()$ is introduced in Section 1.1. Clearly, the integral

$$\tilde{\psi}(A) := [\frac{1}{2\pi}\int_{-\infty}^{\infty} \phi^2(iy)dy]^{1/2}$$

converges and $\psi(A) \le \tilde{\psi}(A)$. Below we will give simple estimates for $\tilde{\psi}(A)$. The previous theorem implies

Corollary 5.3.2 *Let matrix A be Hurwitzian and the conditions (1.2) and*

$$\tilde{b}(A) := \sup_{|y|\le 2\|A\|} \phi(iy) < \frac{1}{\nu}$$

hold. Then the zero solution to equation (1.1) is asymptotically stable. Moreover, with the notation

$$\tilde{\eta}(A) := \frac{\sqrt{2(\|A\| + \nu)}\tilde{\psi}(A)}{1 - \nu\tilde{b}(A)}$$

any vector $x_0 \in \mathbf{C}^n$, satisfying the inequality

$$\|x_0\|\tilde{\eta}(A) < r$$

belong to the region of attraction of the zero solution to equation (1.1). Besides, any solution x of problem (1.1), (1.3) satisfies the inequalities $\sup_{t\ge 0}\|x(t)\| \le \|x_0\|\tilde{\eta}(A)$ and

$$\|x\|_{L^2} \le \frac{\|x_0\|\tilde{\psi}(A)}{1 - \nu\tilde{b}(A)}.$$

It is simple to check that if the spectrum is real, then

$$|det\,(A - iyI)| \ge |\alpha - iy|^n = (\alpha^2 + y^2)^{n/2},$$

where $\alpha = \alpha(A) = \max\ Re\ \lambda_k(A)$. Thus,

$$\psi^2(A) \le \frac{\gamma_n^2}{2\pi}\int_{-\infty}^{\infty} \frac{N^{2n-2}(A - iIy)}{(\alpha^2 + y^2)^n}dy,$$

where

$$\gamma_n = \frac{1}{(n-1)^{(n-1)/2}}.$$

But

$$N(A - Iiy) \le \sqrt{n}|y| + N(A).$$

Therefore

$$\tilde{\psi}^2(A) \le \frac{\gamma_n^2}{\pi} \int_0^\infty \frac{(\sqrt{n}y + N(A))^{2n-2} dy}{(\alpha^2 + y^2)^{2n}}.$$

This integral is simple calculated.

5.3.2 Proof of Theorem 5.3.1

Lemma 5.3.3 *The relation*

$$\sup_{y \in R^1} \|(A - Iiy)^{-1}\| = \sup_{|y| \le 2\|A\|} \|(A - Iiy)^{-1}\| = b(A)$$

is valid.

Proof: Clearly,

$$\sup_{y \in R^1} \|(A - Iiy)^{-1}\| \ge \|A^{-1}\| \ge \frac{1}{\|A\|}.$$

But

$$\|(A - Iiy)^{-1}\| \le \frac{1}{|y| - \|A\|} \le \frac{1}{\|A\|} \quad (|y| \ge 2\|A\|).$$

So the supremum is attained on $[-2\|A\||, 2\|A\||]$. As claimed. □

Furthermore, recall that the space $L^2 = L^2(R_+, \mathbf{C}^n)$ is introduced in Section 1.13. Due to the Laplace transform, Parseval equality and previous lemma, we get

$$\int_0^\infty \|\int_0^t e^{A(t-s)} f(s) ds\|^2 dt = \frac{1}{2\pi} \int_{-\infty}^\infty \|(A - Iiy)^{-1} \tilde{f}(iy)\|^2 dy$$

$$\le b^2(A) \frac{1}{2\pi} \int_{-\infty}^\infty \|\tilde{f}(iy)\|^2 dy = b^2(A) \|f\|_{L^2}^2, \tag{3.5}$$

where $\tilde{f}$ is the Laplace transform to $f \in L^2$. Moreover,

$$\int_0^\infty \|e^{A(s)}\|^2 ds = \frac{1}{2\pi} \int_{-\infty}^\infty \|(A - Iiy)^{-1}\|^2 dy = \psi^2(A). \tag{3.6}$$

First, let condition (1.2) hold with $r = \infty$. Then for any $h \in L^2$, (1.2) implies

$$\|F(h,t)\|_{L^2} \le \nu \|h\|_{L^2}.$$

Rewrite (1.1) as

$$x(t) = e^{At}x_0 + \int_0^t e^{A(t-s)}F(x(s), s)ds. \tag{3.7}$$

Due to (3.6) and (3.5)

$$\|x\|_{L^2} \leq \psi(A)\|x_0\| + \nu b(A)\|x\|_{L^2}.$$

Thus (3.1) yields

$$\|x\|_{L^2} \leq \psi(A)\|x_0\|(1 - \nu b(A))^{-1}.$$

Moreover, from (1.1) it follows that

$$\|\dot{x}\|_{L^2} \leq \|A\|\|x\|_{L^2} + \nu\|x\|_{L^2}.$$

Let us use the inequality

$$v^2(t) \leq 2[\int_t^\infty v^2(s)ds]^{1/2}[\int_t^\infty (\dot{v}(s))^2 ds]^{1/2} \quad (v, \dot{v} \in L^2[0, \infty))$$

(see Section 1.13). Hence, we get

$$\sup_{t\geq 0}\|x(t)\| \leq \sqrt{2(\|A\| + \nu)}\|x\|_{L^2} \leq$$

$$\sqrt{2(\|A\| + \nu)}\psi(A)\|x_0\|(1 - \nu b(A))^{-1} = \eta(A)\|x_0\|.$$

So in the case $r = \infty$ the theorem is proved.

Now let $r < \infty$. Due to (3.1), for a small enough $t_0 > 0$ a solution w of problem (1.1), (1.2) lies in $\Omega(r)$ for all $t \leq t_0$. Applying our above reasonings, we get the inequality

$$\sup_{0\leq t\leq t_0} \|x(t)\| \leq \eta(A)\|x_0\|.$$

But under (3.2) this inequality can be extended from $[0, t_0]$ to $[0, \infty)$. This proves the theorem. □

5.4 Global Stability Under Matrix Conditions

Consider system (1.1) in the form

$$\frac{dx_j}{dt} = \sum_{k=1}^n a_{jk}x_k + F_j(x, t)$$

$$(x = (x_k(t)) \in \mathbf{C}^n;\ t > 0;\ j = 1, ..., n) \tag{4.1}$$

under the conditions

$$|F_j(h,t)| \leq \sum_{k=1}^{n} q_{jk}|h_k|\ (h = (h_k) \in \mathbf{C}^n;\ t \geq 0;\ j = 1, ..., n)$$

where q_{jk} $(j, k = 1, ..., n)$ are nonnegative constants. We will write these conditions in the form

$$|F(x,t)| \leq Q|h| \quad (h \in \mathbf{C}^n;\ t \geq 0) \tag{4.2}$$

where

$$Q = (q_{jk})_{j,k=1}^{n}$$

is a constant matrix. Inequality (4.2) permits us to obtain estimates for each coordinate of a solution and to take into account information about each element of the matrix more completely than under the customary condition (1.2), although applications of condition (4.2) require more calculations than condition (1.2).

Furthermore, let $m_{ij}(\lambda)$ be the cofactor of the element $\delta_{ij}\lambda - a_{ij}$ of the matrix $\lambda I - A$. Recall that $\delta_{ij} = 0,\ j \neq i$ and $\delta_{jj} = 1$. Assume that after division by the coinciding zeros we obtain the equality

$$\frac{m_{ij}(\lambda)}{det(I\lambda - A)} = \frac{\mu_{ij}(\lambda)}{d_{ij}(\lambda)} \quad (i, j = 1, ...n),$$

where $d_{ij}(\lambda)$ and $\mu_{ij}(\lambda, t)$ are polynomials with respect to λ, and the degree of $d_{ij}(\lambda)$ is equal to n_{ij}. Let $co(A)$ be the convex hull of all eigenvalues of matrix A. Put

$$b_{ji}(k) = \frac{1}{k!(n_{ij} - 1 - k)!} \sup_{\lambda \in co(A)} |\mu_{ij}^{(n_{ij}-k-1)}(\lambda)| \tag{4.3}$$

$$(k = 0, ..., n-1).$$

Since $(-m)! = 0$ for any natural m, the equalities

$$b_{ij}(k) = 0 \text{ for } k = n_{ij}, ..., n-1$$

are valid if $n_{ij} < n - 1$. Finally, denote

$$B_k = (b_{ij}(k))_{j,i=1}^{n}.$$

Theorem 5.4.1 *If F satisfies condition (4.2), then any solution $x(t)$ of system (4.1) subordinates the inequality*

$$|x(t)| \leq exp[(\alpha(A) + z_0)t]C(t)|x(0)|\ (t \geq 0),$$

where $C(t)$ is a non-negative polynomial matrix and z_0 is the extreme right-hand (positive) root of the polynomial

$$det(I\lambda^n - \sum_{k=0}^{n-1} \lambda^{n-k-1} B_k Q). \tag{4.4}$$

Recall that I is the unit matrix.

Corollary 5.4.2 *Under condition (4.2), let the matrix $A + z_0 I$ be a Hurwitz one. Then system (4.1) is globally exponentially stable.*

5.5 Proof of Theorem 5.4.1

Condition (4.2) implies

$$|x(t)| \leq |e^{At}x(0)| + \int_0^t |e^{A(t-\tau)}|Q|x(\tau)|d\tau.$$

According to Corollary 1.12.2,

$$|e^{At}| \leq e^{\alpha_0 t}\Psi(t) \ (\tau, t \geq 0),$$

where

$$\Psi(t) = \sum_{k=0}^{n-1} t^k B_k.$$

Consequently,

$$|x(t)|e^{-\alpha_0 t} \leq \Psi(t)|x(0)| + \int_0^t \Psi(t-\tau)Qe^{-\alpha_0 \tau}|x(\tau)|d\tau.$$

It follows from Lemma 1.7.1 that

$$|x(t)|e^{-\alpha_0 t} \leq \eta(t), \tag{5.1}$$

where $\eta(t)$ is the solution of the equation

$$\eta(t) = \Psi(t)|x(0)| + \int_0^t \Psi(t-\tau)Q\eta(\tau)d\tau.$$

To solve that equation we apply the Laplace transformation. Let λ be the dual variable, $\overline{\eta}(\lambda)$ and $\overline{\Psi}(\lambda)$ be the Laplace transforms of $\eta(t)$ and $\Psi(t)$. Using the fact that the transform of a convolution is the product of the functions in the convolution, we obtain

$$\overline{\eta}(\lambda) = \overline{\Psi}(\lambda)|x(0)| + \overline{\Psi}(\lambda)Q\overline{\eta}(\lambda).$$

By the inverse Laplace transform,

$$\eta(t) = \frac{1}{2\pi i}\int_{Re\lambda > z_0} e^{t\lambda}(I - \overline{\Psi}(\lambda)Q)^{-1}\overline{\Psi}(\lambda)|x(0)|d\lambda. \tag{5.2}$$

Obviously,

$$\overline{\Psi}(\lambda) = \sum_{k=0}^{n-1} \frac{k!}{\lambda^{k+1}} B_k.$$

By (5.2) we conclude that

$$\eta(t) = \frac{1}{2\pi i} \int_{Re\lambda > z_0} e^{t\lambda} T^{-1}(\lambda) \sum_{k=0}^{n-1} k! \lambda^{n-k} B_k \, d\lambda \, |x_0| \tag{5.3}$$

with the notation

$$T(\lambda) = I\lambda^n - \sum_{k=0}^{n-1} k! \lambda^{n-k-1} B_k Q.$$

Take into account that

$$T^{-1}(\lambda) = \frac{T_1(\lambda)}{det\, T(\lambda)},$$

where $T_1(\lambda)$ is a polynomial matrix. Clearly, z_0 is a positive number. Now according to the residue theorem, the assertion of the theorem follows from inequality (5.1). □

5.6 Region of Attraction of One-Contour Systems

Let us consider the real scalar equation

$$P(D)y = F(y,t) \;\; (t \geq 0, \; D \equiv d/dt), \tag{6.1}$$

where $y = y(t)$ and

$$P(\lambda) = \lambda^n + a_1 \lambda^{n-1} + ... + a_n$$

is a real Hurwitz polynomial. Besides, $F(y,t)$ is a real continuous scalar-valued function defined on

$$[-r, r] \times [0, \infty)$$

with a positive number $r \leq \infty$. It is assumed that there is a positive constant $q < \infty$, such that

$$|F(x,t)| \leq q|x| \;\; (x \in [-r, r], t \geq 0). \tag{6.2}$$

Let $\lambda_1, ..., \lambda_n$ be the roots of polynomial $P(\lambda)$. Denote

$$\alpha = \max_{1 \leq k \leq n} Re\, \lambda_k,$$

$$\Lambda = \max_{1 \leq k \leq n} |\lambda_k|,$$

and

$$\zeta_j = \sum_{k=0}^{n-1} \frac{j! \Lambda^{j-k} (n-k-1)^{n-k-1}}{|e\alpha|^{n-k-1} (j-k)! k! (n-k-1)!}$$

$$(j = 0, 1, 2, ...).$$

Besides, for the initial values

$$y^{(j)}(0) = y_j \ (j = 0, ..., n-1) \tag{6.3}$$

set

$$w_i(P, 0) \equiv \sum_{j=0}^{n-1} |y_{n-j-1}| \sum_{k=0}^{j} a_k \zeta_{j-k+i} \ \ (i = 0, ..., n-1).$$

Theorem 5.6.1 *Let the conditions (6.2) and*

$$q < |\alpha|^n \tag{6.4}$$

be fulfilled. Then the zero solution of (6.1) is asymptotically stable, and all initial values satisfying the inequality

$$M_0 \equiv \frac{|\alpha|^n w_0(P, 0)}{|\alpha|^n - q} < r \tag{6.5}$$

belong to the region of attraction of the zero solution of (6.1). Besides, any solution $y(t)$ of problem (6.1), (6.3) subordinates the estimate

$$\sup_{t \geq 0} |y(t)| \leq M_0 \ \ (j = 0, 1, ..., n-1). \tag{6.6}$$

The proof of this theorem is presented in the next section.

Let

$$A = \begin{pmatrix} -a_1 & \dots & -a_{n-1} & -a_n \\ 1 & \dots & 0 & 0 \\ . & \dots & . & . \\ 0 & \dots & 1 & 0 \end{pmatrix}.$$

be the matrix of the linear part of equation (6.1). Assume that $A + q^{1/n} I$ is a Hurwitz matrix. Here I is the unit matrix. Then clearly $\alpha + q^{1/n} < 0$. That is, (6.4) holds. Now Theorem 5.6.1 implies the following result

Corollary 5.6.2 *Let condition (6.2) hold and $A + q^{1/n} I$ be a Hurwitz matrix. Then the zero solution of (6.1) is asymptotically stable. Moreover, inequality (6.4) is valid, and all initial values satisfying (6.5) belong to the region of attraction of the zero solution of (6.1). In addition, estimate (6.6) is true.*

5.7 Proof of Theorem 5.6.1

$$K(t) = \frac{1}{2\pi i}\int_C \frac{e^{t\lambda}d\lambda}{P(\lambda)},$$

where C is a smooth contour encompassing all the roots of $P(\lambda)$.

Lemma 5.7.1 *For each natural $j = 0, 1, 2, ...$, the inequality*

$$|K^{(j)}(t)| \leq e^{\alpha t}\eta_j(t) \ \ (t \geq 0)$$

is true, where

$$\eta_j(t) = \sum_{k=0}^{n-1} \frac{j! \Lambda^{j-k} t^{n-k-1}}{(j-k)!(n-k-1)!k!}.$$

Proof: We can write down

$$K^{(j)}(t) = \frac{1}{2\pi i}\int_C \frac{\lambda^j e^{t\lambda}}{\prod_{k=1}^n (\lambda - \lambda_k)} d\lambda,$$

According to Lemma 1.11.1,

$$|K^{(j)}(t)| \leq \frac{1}{(n-1)!} \sup_{\lambda \in co(P)} |\frac{d^{n-1}}{d\lambda^{n-1}} e^{\lambda t}\lambda^j|,$$

where $co(P)$ is the closed convex hull of the numbers $\lambda_1, ..., \lambda_n$. This clearly forces

$$|K^{(j)}(t)| \leq \sum_{k=0}^{n-1} \frac{j! t^{n-k-1}}{k!(n-k-1)!(j-k)!} \sup_{\lambda \in co(P)} |e^{\lambda t}\lambda^{j-k}| \leq$$

$$e^{\alpha t}\sum_{k=0}^{n-1} \frac{j! t^{n-k-1}}{k!(n-k-1)!(j-k)!} \sup_{\lambda \in co(P)} |\lambda^{j-k}|.$$

Since

$$\sup_{\lambda \in co(P)} |\lambda| = \Lambda,$$

this is precisely the assertion of the lemma. □

Corollary 5.7.2 *For each $j = 0, 1, 2, ...$, the inequality*

$$\max_{t \geq 0} |K^{(j)}(t)| \leq \zeta_j$$

is true.

Indeed,

$$\max_{t\geq 0} e^{\alpha t} t^k = k^k |\alpha|^{-k} e^{-k} \quad (k = 1, 2, ...).$$

Hence,

$$\max_{t\geq 0} e^{\alpha t} \eta_j(t) \leq \sum_{k=0}^{n-1} \frac{j! \Lambda^{j-k} (n-k-1)^{n-k-1}}{|e\alpha|^{n-k-1} (j-k)! k! (n-k-1)!} = \zeta_j. \tag{7.1}$$

Thus, the result follows from Lemma 5.7.1.

Corollary 5.7.3 *For each $j = 0, ..., n-1$ the following inequality is valid:*

$$\int_0^\infty |K^{(j)}(t)| dt \leq \frac{(\Lambda + |\alpha|)^j}{|\alpha|^n}.$$

In fact, due to Lemma 5.7.1

$$\int_0^\infty |K^{(j)}(t)| dt \leq \sum_{k=0}^{n-1} \frac{j! \Lambda^{j-k}}{k!(n-k-1)!(j-k)!} \int_0^\infty e^{\alpha t} t^{n-k-1} dt =$$

$$\sum_{k=0}^{j} \frac{j! \Lambda^{j-k}}{k! |\alpha|^{n-k} (j-k)!} = |\alpha|^{-n} (\Lambda + |\alpha|)^j,$$

as claimed.

Lemma 5.7.4 *For a solution $z(t)$ of the equation*

$$P(D) z = 0, \tag{7.2}$$

with the initial conditions

$$z^{(j)}(0) = y_j \quad (j = 0, ..., n-1) \tag{7.3}$$

the following estimate is true:

$$|z^{(m)}(t)| \leq e^{\alpha t} \sum_{i=0}^{n-1} |y_{n-i-1}| \sum_{s=0}^{i} a_s \eta_{i-s+m}(t) \quad (t \geq 0, \ m = 0, 1, ...).$$

Proof: By the Laplace transformation (Dóetch, 1961, formula (14.6)), we have

$$z(t) = \sum_{j=0}^{n-1} z^{(j)}(0) \sum_{k=1}^{n-j} a_{k-1} K^{(n-k-j)}(t).$$

This implies

$$z^{(m)}(t) = \sum_{j=0}^{n-1} z^{(j)}(0) \sum_{k=1}^{n-j} a_{k-1} K^{(n-k-j+m)}(t) =$$

$$\sum_{i=0}^{n-1} z^{(n-i-1)}(0) \sum_{k=1}^{i+1} a_{k-1} K^{(i+1-k+m)}(t) = \sum_{i=0}^{n-1} z^{(n-i-1)}(0) \sum_{s=0}^{i} a_s K^{(i-s+m)}(t).$$

Now the required result follows from Lemma 5.7.1. □

Lemma 5.7.4 and (7.1) yield

Corollary 5.7.5 *For a solution $z(t)$ of problem (7.2), (7.3), the estimate*

$$max_{t\geq 0}|z^{(m)}(t)| \leq w_m(P,0) \ \ (m = 0, 1, 2, ...)$$

is true.

Proof of Theorem 5.6.1: The equation (6.1) is equivalent to the following one:

$$y(t) = z(t) + \int_0^t K(t-s)F(y,s)ds \tag{7.4}$$

where $z(t)$ is a solution of the linear problem (7.2), (7.3). Since (6.5) implies that $|y(0)| < r$, there is a sufficiently small t_0 such that

$$|y(t)| \leq r \text{ for } 0 \leq t \leq t_0.$$

We conclude from (7.4) that

$$|y(t)| \leq |z(t)| + \int_0^t |K(t-s)|q|y(s)|ds, \ 0 \leq t \leq t_0.$$

Hence,

$$\max_{t\leq t_0} |y(t)| \leq max_{t\geq 0}|z(t)| + q \max_{t\leq t_0} |y(t)| \int_0^\infty |K(s)|ds.$$

Due to Corollaries 5.7.5 and 5.7.3 we thus get

$$\max_{t\leq t_0} |y(t)| \leq \max_{t\geq 0} |z(t)| + q|\alpha|^{-n} \leq w_0(P,0) + q|\alpha|^{-n} \max_{t\leq t_0} |y(t)|.$$

But $q|\alpha|^{-n} < 1$. Therefore,

$$\max_{0\leq t\leq t_0} |y(t)| \leq \theta_0 \equiv (1 - q|\alpha|^{-n})^{-1} w_0(P,0) \ \ (0 \leq t \leq t_0). \tag{7.5}$$

According to (6.5) we can extend this estimate to all $t \geq 0$. So we have obtained estimate (6.6). Thus, the stability of the zero solution has been proved. The asymptotic stability can be obtained by a small perturbation of equation (6.1). The detailed verification of this fact is left to the reader. □

6. The Aizerman Problem

6.1 Absolute Stability

Consider the equation

$$\dot{y} = Ay + b\, f(s,t) \ \ (s = cy,\ t \geq 0), \tag{1.1}$$

where A is a real constant Hurwitz $n \times n$-matrix, b is a real column, c is a real row, f maps $\mathbf{R}^1 \times [0, \infty)$ into $\mathbf{R}^1$ with the property

$$|f(s,t)| \leq q|s| \text{ for all } s \in \mathbf{R}^1 \text{ and } t \geq 0. \tag{1.2}$$

Definition 6.1.1 *We will say that the zero solution of system (1.1) is absolutely exponentially stable in the class of nonlinearities (1.2) if there are constants $M, \epsilon > 0$ which do not depend on a concrete form of f (but which depend on q) such that*

$$|cy(t)| \leq M\ exp(-\epsilon t)\ \|y(0)\| \ \ (t \geq 0)$$

for any solution $y(t)$ of (1.1).

Introduce the linear equation

$$\dot{y} = Ay + q_1 bcy \tag{1.3}$$

In 1949 M. A. Aizerman conjectured the following hypothesis: under the condition $f(s,t) \equiv f(s)$, for the absolute stability of the zero solution of (1.1) in the class of nonlinearities satisfying the condition

$$0 \leq f(s)/s \leq q \text{ for all } s \in \mathbf{R}^1, s \neq 0 \text{ and } t \geq 0$$

it is necessary and sufficient that (1.3) be asymptotically stable for any $q_1 \in [0, q]$ (Aizerman, 1949). This hypothesis caused great interest among the specialists. Counterexamples were set up that demonstrated it was not, in general, true (see (Naredra and Taylor, 1973), (Reissig et al., 1974), (Willems, 1971), and references therein).

Therefore, the following problem arose: to find the class of systems that satisfy Aizerman's hypothesis. To formulate the relevant theorem, let us introduce the transfer function $W(\lambda)$ of the linear part of system (1.1):

$$W(\lambda) = c(\lambda I - A)^{-1}b = \frac{L(\lambda)}{P(\lambda)} \quad (\lambda \in \mathbf{C}).$$

Here

$$P(\lambda) = \lambda^n + a_1\lambda^{n-1} + ... + a_n \quad (a_1, ..., a_n > 0)$$

and

$$L(\lambda) = \sum_{k=0}^{m} b_k \lambda^{m-k} \quad (m < n)$$

are polynomials. It is assumed that $P(\lambda)$ and $L(\lambda)$ have no coinciding roots. In addition, let

$$K(t) := \frac{1}{2\pi} \int_{-\infty}^{\infty} exp[i\omega t] W(i\omega)\, d\omega$$

be the impulse (Green) function.

Theorem 6.1.2 *Let the condition*

$$K(t) \geq 0 \text{ for all } t \geq 0 \tag{1.4}$$

be fulfilled. Then for the absolute exponential stability of the zero solution of (1.1) in the class of nonlinearities (1.2), it is necessary and sufficient that the polynomial $P(\lambda) - qL(\lambda)$ be Hurwitzian.

That theorem is proved in the next section

Clearly, Theorem 6.1.2 singles out one of the classes of linear parts of systems that satisfy the conjecture of M.A. Aizerman. In the next section we will prove

Lemma 6.1.3 *Let condition (1.4) be fulfilled. Then the polynomial $P(\lambda) - qL(\lambda)$ is a Hurwitz one, if and only if*

$$qW(0) < 1. \tag{1.5}$$

Now Theorem 6.1.2 implies

Corollary 6.1.4 *Let condition (1.4) be fulfilled. Then for the absolute exponential stability of the zero solution of (1.1) in the class of nonlinearities (1.2), it is necessary and sufficient that condition (1.5) holds.*

Furthermore, immediately from Lemma 1.11.2 it follows

Lemma 6.1.5 *Let all the roots of polynomial $P(\lambda)$ belong to a real segment $[a, b]$ $(b < 0)$ and*

$$L^{(k)}(z) \geq 0 \quad (k = 0, ..., deg\ L(\lambda);\ a \leq z \leq b). \tag{1.6}$$

Then $K(t) \geq 0$ $(t \geq 0)$.

This result, Theorem 6.1.2 and Lemma 6.1.3 imply

Corollary 6.1.6 *Let all the roots of $P(\lambda)$ belong to a real segment $[a,b]$ and condition (1.6) hold. Then for the absolute exponential stability of the zero solution of (1.1) in the class of nonlinearities (1.2) it is necessary and sufficient that the polynomial $P(\lambda) - qL(\lambda)$ be Hurwitzian. That is,*

$$a_n = P(0) > qL(0) = qb_m.$$

In particular, let

$$L(\lambda) \equiv 1$$

and all the roots of $P(\lambda)$ be real. Then for the absolute exponential stability of the zero solution of (1.1) in the class of nonlinearities (1.2) it is necessary and sufficient that $a_n > q$.

In the next section we also prove the following result

Lemma 6.1.7 *Let a real Hurwitz polynomial $P_3(\lambda) = \lambda^3 + a_1\lambda^2 + a_2\lambda + a_3$ have a pair of complex conjugate roots: $-\gamma \pm i\omega$, and a real root $-z_0$ $(z_0, \gamma, \omega > 0)$. Then the function*

$$K_3(t) \equiv \frac{1}{2\pi}\int_{-\infty}^{\infty} exp(ity)P_3^{-1}(iy)dy$$

is positive, provided $\gamma > z_0$.

Since the Laplace transform of a convolution is a product of the Laplace transforms, we can extend this result to an arbitrary $n > 3$. The previous lemma, and Lemma 6.1.3 imply

Corollary 6.1.8 *Let a real Hurwitz polynomial $P_3(\lambda) = \lambda^3 + a_1\lambda^2 + a_2\lambda + a_3$ have a pair of complex conjugate roots: $-\gamma \pm i\omega$, and a real root $-z_0 > -\gamma$ $(z_0, \gamma, \omega > 0)$. Then for the absolute exponential stability of the zero solution of (1.1) with $W(\lambda) = P_3^{-1}(\lambda)$ in the class of nonlinearities (1.2) it is necessary and sufficient that $a_3 > q$.*

6.2 Proofs

Proof of Theorem 6.1.2: First, we will prove the sufficiency of the conditions. Note that

$$K(t) = c\exp[At]b.$$

Furthermore, equation (1.1) is equivalent to following one:

$$y(t) = exp[At]y(0) + \int_0^t exp[A(t-\tau)]bf(s(\tau),\tau)d\tau \ \ (s(t) = cy(t)).$$

Multiplying this equation by c, we have

$$s(t) = z(t) + \int_0^t K(t-\tau) f(s(\tau), \tau) d\tau$$

with

$$z(t) = c\, exp[At] y(0).$$

Taking into account condition (1.2), we arrive at the inequality

$$|s(t)| \leq |z(t)| + \int_0^t K(t-\tau) q |s(\tau)| d\tau$$

From Lemma 1.7.1 it follows that

$$|s(t)| \leq \eta(t), \tag{2.1}$$

where η is the solution of the equation

$$\eta(t) = |z(t)| + \int_0^t K(t-\tau) q \eta(\tau) d\tau.$$

We will solve this equation by the Laplace transformation. Since the transform of the convolution is equal to the product of the transforms, after straightforward calculations, we get the equation

$$\overline{\eta}(\lambda) = g(\lambda) + P^{-1}(\lambda) L(\lambda) q \overline{\eta}(\lambda),$$

where $\overline{\eta}(\lambda)$ and $g(\lambda)$ are the Laplace transforms of $\eta(t)$ and of $|z(t)|$, respectively, λ is the dual variable. Thus we can write down

$$\eta(t) = \frac{1}{2\pi i} \int_{-i\infty + c_0}^{i\infty + c_0} \exp[\lambda t](P(\lambda) - L(\lambda) q)^{-1} P(\lambda) g(\lambda) d\lambda \ \ (c_0 = const).$$

It is easy to see that

$$|z(t)| \leq exp[\alpha(A)t] G(t),$$

where $G(t)$ is a polynomial. Recall that $\alpha(A)$ is the real part of the extreme right-hand eigenvalue of A. Thus, thanks the residue theorem

$$\eta(t) \leq \exp(m_1 t) G_1(t) \text{ for all } t \geq 0,$$

where $G_1(t)$ is a polynomial, and $m_1 = max\{\alpha(A), \beta\}$. Here β is the real part of the extreme right-hand zero of the polynomial $det(P(\lambda) - L(\lambda) q)$. This and (2.1) prove the sufficiency of conditions.

To prove the necessity of the conditions put $f(s,t) = qs$. Then (1.1) take the form

$$\dot{y} = Ay + bqs \ \ (t \geq 0, s = cy).$$

Hence,

$$\lambda \overline{y}(\lambda) - y(0) = A\overline{y} + bqc\overline{y}(\lambda),$$

where $\overline{y}(\lambda)$ is the Laplace transform of $y(t)$. Consequently,

$$c\overline{y}(\lambda) = -c(A - I\lambda)^{-1}(y(0) + bqc\overline{y}(\lambda)) = h(\lambda) + W(\lambda)bqc\overline{y}(\lambda)$$

where $h(\lambda) = -c(A - I\lambda)^{-1}y(0)$. Therefore,

$$c\overline{y}(\lambda) = h(\lambda) + P^{-1}(\lambda)L(\lambda)qc\overline{y}(\lambda)$$

and

$$c\overline{y}(\lambda) = (P(\lambda) - L(\lambda)q)^{-1}P(\lambda)h(\lambda).$$

Hence the required result follows. □

Proof of Lemma 6.1.3: Since $P(\lambda)$ is a Hurwitz polynomial, according to the Rauscher theorem we need only to check that $|P(i\omega)| > q|L(i\omega)|$ or $q|W(i\omega)| < 1$ $(\omega \in \mathbf{R})$, but due to (1.4) this inequality follows from the equality

$$\max_{s\in\mathbf{R}} |W(is)| = W(0) > 0,$$

since

$$|W(is)| = |\int_0^\infty e^{-its}K(t)dt| \leq \int_0^\infty |K(t)|dt$$
$$= \int_0^\infty K(t)dt = W(0) \ (s \in \mathbf{R}).$$

□

Proof of Lemma 6.1.7: Taking into account that $K_3(t)$ is a solution of the equation

$$P_3(D)x(t) = 0,$$

with the initial condition $K_3(0) = \dot{K}_3(0) = 0$, $\ddot{K}(0) = 1$ (Döetsch, 1961, Section 12), we get

$$K_3(t) = c[e^{-z_0 t} - e^{-\gamma t}(cos(\omega t) + b\ sin(\omega t))] \ (c = const > 0)$$

with

$$b = (\gamma - z_0)\omega^{-1} > 0.$$

Consider the equation $K_3(t) = 0$. Multiplying it by $e^{\gamma t}$ and substituting $s = \omega t$, we have the equation

$$e^{bs} = cos\ s + b\ sin\ s.$$

By virtue of the Taylor series that equation can be rewritten as

$$e^{bs} = 1 + bs + g(s) = cos\ s + b\ sin\ s \ \ (g(s) > 0).$$

Since

$$|cos\ s| \leq 1, |sin\ s| \leq s \ (s \geq 0),$$

one can assert that this equation has no zeros for $s > 0$. Therefore, $K_3(t)$ has no zeros. □

6.3 Systems with Matrix Conditions

Now we consider the system

$$\dot{y} = Ay + Bf(s,t), \ (t \geq 0, s = Cy) \tag{3.1}$$

where A is a real Hurwitz $n \times n$-matrix,

$$B : \mathbf{R}^m \to \mathbf{R}^n \text{ and } C : \mathbf{R}^n \to \mathbf{R}^m$$

are linear operators also, and $m < n$. Besides, $F(s,t)$ maps $\mathbf{R}^m \times [0,\infty)$ into $\mathbf{R}^m$ with the property

$$|F(x,t)| \leq Q|x| \ (x \in \mathbf{R}^m, \ t \geq 0). \tag{3.2}$$

Here Q is a non-negative $m \times m$-matrix ; $|F(x,t)|$ and $|x|$ are vectors whose coordinates are the moduli of the vectors $F(x,t)$ and x, respectively.

Let $W(\lambda)$ be the transfer matrix of the linear part of the system from the input F to the output y:

$$W(\lambda) = C(\lambda I - A)^{-1}B.$$

It is simple to check that $W(\lambda) = P^{-1}(\lambda)L(\lambda)$, where $L(\lambda)$ and $P(\lambda)$ are matrices whose elements are polynomials in λ. So each element $W_{jk}(\lambda)$ of $W(\lambda)$ $(j,k = 1,...,m)$ is a rational quotient function and the degree of the numerator of $W_{jk}(\lambda)$ is less than the degree of the denominator. It is assumed that $W(\lambda)$ is non-degenerate, i.e. the poles of this matrix are not canceled by the zeros. In particular, (3.1) can take the form

$$P(D)y = L(D)F(y,t) \ \ (t \geq 0; \ D = d/dt).$$

Let

$$K(t) = \frac{1}{2\pi} \int_{-i\infty}^{i\infty} exp[i\omega t] W(i\omega) d\omega \tag{3.3}$$

be the impulse (Green) matrix of the linear part of the system.

We will say that the zero solution of the system (3.1) is *absolutely exponentially stable in the class of nonlinearities (3.2)* if there are scalar constants $N, \epsilon > 0$ independent of the form of F, such that

$$\|Cy(t)\|_{R^m} \leq N exp(-\epsilon t)\|y(0)\|_{R^n} \text{ for all } t \geq 0$$

and any solution $y(t)$ of system (3.1). Here $\|.\|_{R^n}$ and $\|.\|_{R^m}$ are arbitrary norms in $\mathbf{R}^n$ and $\mathbf{R}^m$, respectively.

Theorem 6.3.1 *Let $K(t) \geq 0$ for all $t \geq 0$. Then for the absolute exponential stability of the zero solution of equation (3.1) in the class of nonlinearities (3.2) it is necessary and sufficient that the polynomial $det(P(\lambda) - L(\lambda)Q)$ be a Hurwitz one.*

The proof of this theorem is similar to the proof of Theorem 6.1.2. Clearly, Theorem 6.3.1 generalizes Theorem 6.1.2.

Furthermore, each element $W_{jk}(\lambda)$ $(j,k=1,...,m)$ of the transfer function $W(\lambda)$ has the form

$$W_{jk}(\lambda)=\frac{m_{jk}(\lambda)}{p_{jk}(\lambda)}, \tag{3.4}$$

where $p_{jk}(\lambda)$ and $m_{jk}(\lambda)$ are polynomials in λ. In addition p_{jj} are monic polynomials. That is their coefficients of the largest powers are equal to one.

Lemma 1.11.2 and Theorem 6.3.1 imply

Corollary 6.3.2 *Under (3.4), for all $j,k=1,...,m$, let all the roots of $p_{jk}(\lambda)$ belong to a real segment $[a_{jk},b]$ and suppose that*

$$\frac{d^\nu m_{jk}(\lambda)}{d\lambda^\nu}\geq 0\ (\lambda\in[a_{jk},b_{jk}],\ t\geq 0;\ \nu=0,1,...,deg\ m_{jk}(\lambda)).$$

Then for the absolute exponential stability of the zero solution of (3.1) in the class of nonlinearities (3.2) it is necessary and sufficient that the polynomial $det(P(\lambda)-L(\lambda)Q)$ be a Hurwitz one.

6.4 Examples

Example 6.4.1 *Consider the equation*

$$\frac{d^2x}{dt^2}+a_1\frac{dx}{dt}+a_2x=b_1\phi(x)+\frac{d\phi(x)}{dt}\ (a_1,a_2,b_1=const>0), \tag{4.1}$$

where $\phi(s)$ satisfies the condition

$$|\phi(s)|\leq q|s|\ (s\in\mathbf{R},\ t\geq 0). \tag{4.2}$$

Let the polynomial

$$P(\lambda)=\lambda^2+a_1\lambda+a_2$$

have real roots $\lambda_1\leq\lambda_2<0$. Then under the condition

$$b_1+\lambda_1\geq 0 \tag{4.3}$$

we have

$$L(\lambda)=\lambda+b_1\geq 0,\ \frac{d}{d\lambda}L(\lambda)=1\ (\lambda_1\leq\lambda\leq\lambda_2).$$

By Corollary 6.1.6, the zero solution of equation (4.1) is absolutely exponentially stable in the class of nonlinearities (4.2), provided the conditions (4.3) and $a_2>qb_1$ hold.

Example 6.4.2 *Consider the equation*

$$\frac{d^3x}{dt^3} + a_1\frac{d^2x}{dt^2} + a_2\frac{dx}{dt} + a_3x = b_1\phi(x) + \frac{d\phi(x)}{dt}$$
$$(a_1, a_2, a_3, b_1 = const > 0), \tag{4.4}$$

where $\phi(x)$ satisfies condition (4.2). Let the polynomial

$$P(\lambda) = \lambda^3 + a_1\lambda^2 + a_2\lambda + a_3$$

have real roots $\lambda_1 \le \lambda_2 \le \lambda_3 < 0$. Then under condition (4.3), relations

$$L(\lambda) = \lambda + b_1 \ge 0, \ \frac{d}{d\lambda}L(\lambda) = 1 \ \ (\lambda_1 \le \lambda \le \lambda_3)$$

hold. By Corollary 6.1.6, the zero solution of equation (4.4) is absolutely exponentially stable in the class of nonlinearities (4.2), provided the conditions (4.3) and $a_3 > qb_1$ hold.

Example 6.4.3 *Consider the system*

$$p_{11}(D)x_1 + p_{12}(D)x_2 = \phi_1(x_1, x_2, t),$$
$$p_{22}(D)x_2 = \phi_2(x_1, x_2, t), \tag{4.5}$$

where $p_{jk}(\lambda)$ are polynomials. In addition $p_{11}(\lambda)$ and $p_{22}(\lambda)$ are Hurwitzian monic polynomials. The scalar-valued functions ϕ_1 and ϕ_2 satisfy the conditions

$$|\phi_j(z_1, z_2, t)| \le q_{j1}|z_1| + q_{j2}|z_2| \ (j = 1, 2) \tag{4.6}$$

for all $z_1, z_2 \in \mathbf{R}^1$ and $t \ge 0$. Assume that all the zeros of $p_{11}(\lambda)$ and $p_{22}(\lambda)$ belong to a real segment $[a, b]$ $(a < b < 0)$. In the considered case

$$L(\lambda) = I; \ Q = (q_{jk})_{j,k=1}^2; \ W(\lambda) = (W_{jk}(\lambda))_{j,k=1}^2,$$

where

$$W_{jj}(\lambda) = \frac{1}{p_{jj}}(\lambda) \ (j = 1, 2), \ W_{12}(\lambda) = -\frac{p_{12}(\lambda)}{p_{11}(\lambda)p_{22}(\lambda)}, \ W_{21}(\lambda) = 0.$$

In addition,

$$P(\lambda) = \begin{pmatrix} p_{11}(\lambda) & p_{12}(\lambda) \\ 0 & p_{22}(\lambda) \end{pmatrix}.$$

If

$$p_{12}^{(k)}(\lambda) \le 0 \ (k = 0, 1, ..., deg \ p_{12}(\lambda)) \tag{4.7}$$

for all $\lambda \in [a, b]$, then due to Corollary 6.3.2 for the absolute exponential stability of the zero solution of system (4.5) in the class of nonlinearities (4.6) it is necessary and sufficient that $det(P(\lambda) - Q)$ be a Hurwitz polynomial.

7. Nonlinear Systems with Time-Variant Linear Parts

7.1 Systems with General Linear Parts

Let us consider in $\mathbf{C}^n$ the equation

$$\dot{x} = A(t)x + F(x,t) \quad (t \geq 0), \tag{1.1}$$

where $A(t)$ is a piecewise continuous Hurwitz $n \times n$-matrix, and F maps $\Omega(r) \times [0, \infty)$ into $\mathbf{C}^n$. Recall that $\Omega(r) = \{h \in \mathbf{C}^n : \|h\| \leq r\}$ for a positive number r.

It is assumed that there exists a non-negative piecewise continuous function $\nu(t) = \nu(t, r)$ bounded on $[0, \infty)$, such that

$$\|F(h,t)\| \leq \nu(t)\|h\| \text{ for all } h \in \Omega(r) \text{ and } t \geq 0. \tag{1.2}$$

For any fixed $\tau \geq 0$, let the condition

$$\|exp[A(\tau)t]\| \leq p(t, A(\tau)) \; (t \geq 0) \tag{1.3}$$

hold, where $p(t, A(\tau))$ is a piecewise continuous in t for each $\tau \geq 0$ function, such that

$$\chi := \sup_{t \geq 0} p(t, A(t)) < \infty.$$

Put

$$q(t,s) \equiv \|A(t) - A(s)\| \quad (t, s \geq 0).$$

Theorem 7.1.1 *Let the conditions (1.2), (1.3) and*

$$\theta(A(.), F) \equiv \sup_{t \geq 0} \int_0^t p(t-s, A(t))[q(t,s) + \nu(s)]ds < 1 \tag{1.4}$$

be fulfilled. Then the zero solution to equation (1.1) is stable. Moreover, if

$$\frac{\chi\|x(0)\|}{1 - \theta(A(.), F)} < r, \tag{1.5}$$

then a solution $x(t)$ of (1.1) satisfies the estimate

$$\|x(t)\| \leq \frac{\chi\|x(t)\|}{1 - \theta(A(.), F)} \quad (t \geq 0) \tag{1.6}$$

Proof: Rewrite equation (1.1) in the form

$$dx/dt - A(\tau)x = [A(t) - A(\tau)]x + F(x,t),$$

regarding an arbitrary $\tau \geq 0$ as fixed. This equation is equivalent to the following one:

$$x(t) = exp[A(\tau)t]x(0)+$$

$$\int_0^t exp[(A(\tau)(t-s)][(A(s) - A(\tau))x(s) + F(x(s),s)]ds. \qquad (1.7)$$

Since the solutions continuously depend on the initial vector, the inequality

$$\|x(t)\| < r \ (0 \leq t \leq t_0)$$

is true for a sufficiently small t_0. Due to (1.2) and (1.7), the latter inequality implies the relation

$$\|x(t)\| \leq p(t, A(\tau))\|x(0)\| + \int_0^t p(t-s, A(\tau))[q(\tau,s) + \nu(s)]\|x(s)\|ds \ (t \leq t_0).$$

Taking $\tau = t$, we get

$$\|x(t)\| \leq p(t, A(t))\|x(0)\| + \int_0^t p(t-s, A(t))[q(t,s) + \nu(s)]\|x(s)\|ds \ (t \leq t_0).$$

This relation according to the definitions of $\theta(A(.), F)$ and χ yields

$$\sup_{s \leq t_0} \|x(s)\| < \chi\|x(0)\| + \sup_{s \leq t_0} \|x(s)\|\theta(A(.), F).$$

Consequently, due to (1.4)

$$\sup_{s \leq t_0} \|x(s)\| \leq \chi\|x(0)\|(1 - \theta(A(.), F))^{-1}$$

for a sufficiently small t_0. But condition (1.5) ensures this bound for all $t \geq 0$. That bound provides the (Lyapunov) stability. □

Put

$$\tilde{p}(t, A(s)) = exp[\alpha(A(s))t] \sum_{k=0}^{n-1} \frac{g^k(A(s))t^k}{(k!)^{3/2}}$$

with $\alpha(A(s)) = \max_k Re\ \lambda_k(A(s))$. Assume that

$$\tilde{\chi} := \sup_{t \geq 0} \tilde{p}(t, A(t)) < \infty.$$

Due to Corollary 1.5.3 and the previous theorem we get

Theorem 7.1.2 *Let the conditions (1.2) and*

$$\tilde{\theta}(A(.), F) := \sup_{t \geq 0} \int_0^t \tilde{p}(t-s, A(t))[q(t,s) + \nu(s)]ds < 1$$

be fulfilled. Then the zero solution to equation (1.1) is stable. Moreover, if

$$\frac{\tilde{\chi}\|x(0)\|}{1 - \tilde{\theta}(A(.), F)} < r,$$

then the estimate

$$\|x(t)\| \leq \frac{\tilde{\chi}\|x(0)\|}{1 - \tilde{\theta}(A(.), F)} \quad (t \geq 0)$$

is true for a solution $x(t)$ of (1.1).

7.2 Systems with the Lipschitz Property

7.2.1 Statement of the Result

Assume that the following conditions are satisfied:

$$\|A(t) - A(s)\| \leq q_0|t-s| \quad (t, s \geq 0), \tag{2.1}$$

and

$$\|F(h,t)\| \leq \nu_0\|h\| \quad (h \in \Omega(r), t \geq 0). \tag{2.2}$$

Here q_0 and ν_0 are non-negative constants. In addition, let

$$\|exp[A(\tau)t]\| \leq p_0(t) \ (t, \tau \geq 0), \tag{2.3}$$

where $p_0(t)$ is a piecewise continuous function, such that

$$\chi_0 := \sup_{t \geq 0} p_0(t) < \infty.$$

Then,

$$\int_0^t p_0(t-s)[q_0(t-s) + \nu_0]ds \leq \theta_0(A(.), \nu_0),$$

where

$$\theta_0(A(.), \nu_0) := \int_0^\infty (q_0 t + \nu_0)p(t)dt.$$

Now Theorem 7.1.1 and a small perturbation of equation (1.1) imply

Theorem 7.2.1 *Let the conditions (2.1-2.3) and*

$$\theta_0(A(.), \nu_0) < 1$$

hold. Then the zero solution to equation (1.1) is exponentially stable. In addition, any initial vector x_0 satisfying the inequality

$$\frac{\chi_0 \|x_0\|}{1 - \theta_0(A(.), \nu_0)} < r,$$

belongs to the region of attraction of the zero solution. Moreover, the corresponding solution $x(t)$ of equation (1.1) is subject to the estimate

$$\|x(t)\| \leq \frac{\chi_0 \|x_0\|}{1 - \theta_0(A(.), \nu_0)} \quad (t \geq 0).$$

Furthermore, let $g(.)$ and α be defined as in Sections 1.5 and 1.3, respectively, and

$$v := \sup_{t \geq 0} g(A(t)) < \infty, \text{ and } \rho := -\sup_{t \geq 0} \alpha(A(t)) > 0. \tag{2.4}$$

Put

$$\tilde{\chi}_0 := \sup_{t \geq 0} e^{-\rho t} \sum_{j=0}^{n-1} \frac{v^j}{(j!)^{3/2}}.$$

Below we will prove that the previous theorem implies

Corollary 7.2.2 *Let the conditions (2.1), (2.2), (2.4) and*

$$\Gamma_0(A(.), \nu_0) := \sum_{j=0}^{n-1} \frac{v^j}{\sqrt{j!}} \left(\frac{q_0(j+1)}{\rho^{j+2}} + \frac{\nu_0}{\rho^{j+1}}\right) < 1$$

hold. Then the zero solution to equation (1.1) is exponentially stable. In addition, any initial vector x_0 satisfying the inequality

$$\frac{\tilde{\chi}_0 \|x_0\|}{1 - \Gamma_0(A(.), \nu_0)} < r, \tag{2.5}$$

belongs to the region of attraction of the zero solution. Moreover, the corresponding solution $x(t)$ of equation (1.1) is subject to the estimate

$$\|x(t)\| \leq \frac{\tilde{\chi}_0 \|x_0\|}{1 - \Gamma_0(A(.), \nu_0)} \quad (t \geq 0). \tag{2.6}$$

Furthermore, denote

$$b_j = \frac{j q_0 v^{j-1}}{\sqrt{(j-1)!}} + \frac{\nu_0 v^j}{\sqrt{j!}} \text{ for } j = 1, ..., n-1,$$

$$\text{and } b_0 = \nu_0, \; b_n = q_0 \frac{nv^{n-1}}{\sqrt{(n-1)!}}. \tag{2.7}$$

Let $z(A(.), \nu_0)$ be the extreme right-hand (unique positive and simple) root of the algebraic equation

$$z^{n+1} = \sum_{j=0}^{n} b_j z^{n-j}. \tag{2.8}$$

Theorem 7.2.3 *Let conditions (2.1), (2.2) and (2.4) hold. If, in addition, the matrix $A(t) + (\epsilon + z(A(.), \nu_0))I$ is Hurwitzian for an $\epsilon > 0$ and all $t \geq 0$, then the zero solution to equation (1.1) is exponentially stable, and inequality (2.4) is valid. Moreover, any vector x_0 satisfying (2.5), belongs to the region of attraction of the zero solution and estimate (2.6) is true for the corresponding solution.*

Example 7.2.4 *Consider the scalar equation*

$$\ddot{y} + p(t)\dot{y} + w(t)y = f(t, y, \dot{y}), \tag{2.9}$$

where $p(t)$ and $w(t)$ are non-negative scalar-valued functions with the property

$$|p(t) - p(s)| + |w(t) - w(s)| \leq q_0|t - s| \text{ for all } t, s \geq 0. \tag{2.10}$$

It is assumed that f is a real function defined on $[0, \infty) \times \mathbf{R}^2$ and satisfying the condition

$$f^2(t, y, z) \leq \nu_0^2 \, (y^2 + z^2)$$
$$(t \geq 0; \; z, y \in R^1, \; y^2 + z^2 \leq r^2; \; r \leq \infty). \tag{2.11}$$

Let $A(t)$ be the matrix of the linear part of equation (2.9), and

$$\alpha(A(t)) \leq -\rho < 0 \text{ for all } t \geq 0.$$

Let us suppose that $sup_{t \geq 0} w(t) < \infty$. Then according to Example 1.6.1,

$$g(A(t)) \leq v \equiv 1 + \sup_{t \geq 0} \, w(t).$$

In addition, assume that

$$\Gamma_0(A(.), \nu_0) = \frac{q_0}{\rho^2} + \frac{\nu_0}{\rho} + v(\frac{2q_0}{\rho^3} + \frac{\nu_0}{\rho^2}) < 1.$$

Then by Corollary 7.2.2 the zero solution of (2.9) is exponentially stable. Each vector $x(0) = (y(0), y'(0))$ satisfying (2.5), belongs to the region of attraction of the zero solution. Note that for the considered equation

$$\tilde{\chi}_0 \leq \max_{t \geq 0} exp[-t\rho] \, (1 + vt).$$

7.2.2 Proofs of Corollary 7.2.2 and Theorem 7.2.3

Proof of Corollary 7.2.2: We have

$$\int_0^\infty e^{-\rho t} \sum_{k=0}^{n-1} \frac{v^k t^{k+1}}{(k!)^{3/2}} dt = \theta_1(A),$$

where

$$\theta_1(A) = \sum_{k=0}^{n-1} \frac{(k+1)v^k}{\sqrt{k!}\rho^{k+2}}.$$

Moreover,

$$\int_0^\infty e^{-\rho t} \sum_{k=0}^{n-1} \frac{v^k s^k}{(k!)^{3/2}} ds = \theta_0(A),$$

where

$$\theta_0(A) = \sum_{k=0}^{n-1} \frac{v^k}{\rho^{k+1}\sqrt{k!}}.$$

Now the result is due to Theorem 7.2.1. □

Lemma 7.2.5 *Let $z(A(.), \nu_0)$ be the extreme right-hand root of equation (2.8). Then the inequality*

$$\alpha(A(t)) + z(A(.), \nu_0) < 0 \textit{ for all } t \geq 0$$

implies the inequality $\Gamma_0(A(.), \nu_0) < 1$.

Proof: The hypothesis of the lemma entails the inequality

$$\rho \equiv -\sup \alpha(A(t)) > z(A(.), \nu_0).$$

Dividing (2.8) by z^{n+1}, and taking into account that $z(A(.), \nu_0)$ is its root, we arrive at the inequality

$$1 = \sum_{j=0}^{n} b_j z^{-j-1}(A(.), \nu_0) < \sum_{j=0}^{n} b_j \rho^{-j-1}.$$

But according to (2.7),

$$\Gamma_0(A(.), \nu_0) = \sum_{j=0}^{n} b_j \rho^{-j-1} < 1.$$

As claimed. □

The assertion of Theorem 7.2.3 immediately follows from Corollary 7.2.1 and Lemma 7.2.5.

7.3 Systems with Differentiable Linear Parts

In this section we consider system (1.1) with a real differentiable matrix $A(t) = (a_{jk}(t))_{j,k=1}^n$, which is Hurwitzian for each $t \geq 0$ and uniformly bounded on $[0, \infty)$. We establish stability conditions in terms of the determinant of the variable matrix.

7.3.1 Stability Conditions

Denote

$$\psi(t) = \frac{1}{\pi} \int_{-\infty}^{\infty} \|(A(t) - iI\omega)^{-1}\|^3 d\omega$$

and

$$\zeta(t) = \frac{1}{\pi} \int_{-\infty}^{\infty} \|(A(t) - iI\omega)^{-1}\|^2 d\omega \ \ (t \geq 0).$$

Theorem 7.3.1 *Let the conditions (1.2) and*

$$\|\dot{A}(t)\|\psi(t) + \nu(t)\zeta(t) \leq 1 \ \ (t \geq 0). \tag{3.1}$$

hold. Then the zero solution to equation (1.1) is stable. Moreover, there is a constant $M_0 \geq 1$ such that

$$\|x(t)\| \leq M_0\|x(0)\| \ \ (t \geq 0) \tag{3.2}$$

for any solution $x(t)$ of (1.1), provided

$$M_0\|x(0)\| < r. \tag{3.3}$$

The proof of this theorem is presented in the next subsection. Below we suggest estimates for M_0. We will also prove that Theorem 7.3.1 implies the following result.

Corollary 7.3.2 *Let the conditions (1.2) and*

$$\sup_{t \geq 0} \ \|\dot{A}(t)\|\psi(t) + \nu(t)\zeta(t) < 1 \tag{3.4}$$

hold. Then the zero solution to equation (1.1) is exponentially stable. Moreover, there is a constant $M_0 \geq 1$ such that estimate (3.2) is valid, provided condition (3.3) holds.

Now put

$$\tilde{\psi}(t) := \frac{1}{\pi} \int_{-\infty}^{\infty} \frac{(N^2(A(t)) + n\omega^2)^{3(n-1)/2} d\omega}{|det\ (i\omega I - A(t))|^3 \ \ (n-1)^{3(n-1)/2}}$$

and

$$\tilde{\zeta}(t) := \frac{1}{\pi} \int_{-\infty}^{\infty} \frac{(N^2(A(t)) + n\omega^2)^{n-1} d\omega}{|det\ (i\omega I - A(t))|^2 \ \ (n-1)^{n-1}}.$$

We will check that $\zeta(t) \leq \tilde{\zeta}(t)$ and $\psi(t) \leq \tilde{\psi}(t)$. So the previous theorem implies

Corollary 7.3.3 *Let the conditions (1.2) and*

$$\|\dot{A}(t)\|\tilde{\psi}(t) + \nu(t)\tilde{\zeta}(t) \leq 1 \ \ (t \geq 0)$$

hold. Then the zero solution to equation (1.1) is stable. Moreover, there is a constant $M_0 \geq 1$ such that (3.2) is valid for any solution $x(t)$ of (1.1), provided (3.3) holds.

Note that if all the eigenvalues of the matrix are real, then

$$|det\ (i\omega I - A(t))| \geq (\omega^2 + \alpha^2(t))^{n/2} \ \ (\omega \in \mathbf{R}),$$

where

$$\alpha(t) := \alpha(A(t)) = \max_{k=1,\dots,n} Re\ \lambda_k(A(t))$$

and

$$\lambda_k(A(t)) \ (k = 1, ..., n)$$

are the eigenvalues of matrix $A(t)$ with their multiplicities. Thus,

$$\tilde{\zeta}(t) \leq \frac{1}{\pi} \int_{-\infty}^{\infty} \frac{(N^2(A(t)) + ny^2)^{n-1} dy}{(y^2 + \alpha^2(A))^n \ \ (n-1)^{n-1}}.$$

Take into account that

$$\sum_{k=1}^{n} |\lambda_k(A(t))|^2 \leq N^2(A(t))$$

and

$$|\alpha(t)| \leq \min_k |\lambda_k(A(t))|.$$

So

$$n\alpha^2(t) \leq N^2(A(t))$$

and

$$\tilde{\zeta}(t) \leq \frac{2(2N^2(A(t)))^{n-1}}{\pi} \int_0^{\infty} \frac{ds}{s^2+1}.$$

Consequently, $\tilde{\zeta}(t) \leq \zeta_0(t)$, where

$$\zeta_0(t) = \frac{2^{n-1} N^{2(n-1)}(A(t))}{|\alpha(t)|^{2n-1}(n-1)^{n-1}}.$$

Similarly,

$$\tilde{\psi}(t) \leq \psi_0(t),$$

where

$$\psi_0(t) = \frac{2^{(3n-1)/2} N^{3(n-1)}(A(t))}{(n-1)^{3(n-1)/2} \pi |\alpha(t)|^{3n-1}}.$$

(see Section 3.3). In the case $n = 2$,

$$\psi_0(t) = \frac{4\sqrt{2}N^3(A(t))}{\pi|\alpha(t)|^5} \text{ and } \zeta_0(t) = \frac{2N^2(A(t))}{|\alpha(t)|^3}. \quad (3.5)$$

Corollary 7.3.2 implies

Corollary 7.3.4 *Let all the eigenvalues of $A(t)$ be real and the conditions (1.2), and*

$$\sup_{t\geq 0} \|\dot{A}(t)\|\psi_0(t) + \nu(t)\zeta_0(t) < 1$$

hold. Then the zero solution to equation (1.1) is exponentially stable. Moreover, there is a constant $M_0 \geq 1$, such that estimate (3.2) is valid, provided condition (3.3) holds.

7.3.2 Proof of Theorem 7.3.1

Lemma 7.3.5 *Let condition (1.2) with $r = \infty$ and (3.1) hold. Then there is a constant M_0, such that (3.2) is valid.*

Proof: Recall the Lyapunov theorem (see Section 1.9): if the eigenvalues of a constant matrix A_0 lie in the interior of the left half-plane, then for any positive definite Hermitian matrix H there exists a positive definite Hermitian matrix W_H such that

$$W_H A_0 + A_0^* W_H = -2H.$$

Moreover,

$$W_H = \frac{1}{\pi}\int_{-\infty}^{\infty} (-iI\omega - A_0^*)^{-1} H (iI\omega - A_0)^{-1} d\omega.$$

Now let $A(t)$ depend on t. Put

$$W(t) = \frac{1}{\pi}\int_{-\infty}^{\infty} (-iI\omega - A^*(t))^{-1}(iI\omega - A(t))^{-1} d\omega. \quad (3.6)$$

Then $W(t)$ is a solution of the equation.

$$W(t)A(t) + A^*(t)W(t) = -2I. \quad (3.7)$$

Since $A(.)$ is real,

$$(W(t)A(t)h, h) = -(h, h) \quad (h \in \mathbf{R}^n).$$

Furthermore, multiplying equation (1.1) by $W(t)$ and doing the scalar product, we get

$$(W(t)\dot{x}(t), x(t)) = (W(t)A(t)x(t), x(t)) + (W(t)F(x(t), t), x(t)).$$

But

$$\frac{d}{dt}(W(t)x(t), x(t)) = (W(t)\dot{x}(t), x(t)) + (\dot{W}(t)x(t), x(t)) + (W(t)x(t), \dot{x}(t)) =$$

$$2(W(t)\dot{x}(t), x(t)) + (\dot{x}(t), x(t)).$$

Thus, it can be written

$$\frac{d}{dt}(W(t)x(t), x(t)) = (\dot{W}(t)x(t), x(t)) - 2(x(t), x(t)) + 2(W(t)F(x(t), t), x(t)).$$

Hence the conditions (1.2) and

$$\|\dot{W}(t)\| + 2\nu(t)\|W(t)\| \le 2 \ \ (t \ge 0) \tag{3.8}$$

provides the inequality

$$(W(t)x(t), x(t)) \le (W(0)x(0), x(0)). \tag{3.9}$$

From Lemma 1.10.1 and the uniform boundedness of $A(t)$ it follows that

$$(W(t)h, h) \ge c_0(h, h) \ (c_0 = const > 0, \ t \ge 0, \ h \in \mathbf{R}^n). \tag{3.10}$$

In the next subsection we will establish a lower estimate for c_0. Take into acount tat

$$\dot{W}(t) = \frac{d}{dt} \frac{1}{\pi} \int_{-\infty}^{\infty} (-iI\omega - A^*(t))^{-1}(iI\omega - A(t))^{-1} d\omega =$$

$$\frac{1}{\pi} \int_{-\infty}^{\infty} (-iI\omega - A^*(t))^{-1}[\dot{A}^*(t)(-iI\omega - A^*(t))^{-1} +$$

$$(-iI\omega - A(t))^{-1}\dot{A}(t)](iI\omega - A(t))^{-1} d\omega.$$

Hence,

$$\|\dot{W}(t)\| \le 2\|\dot{A}(t)\| \frac{1}{\pi} \int_{-\infty}^{\infty} \|(iI\omega - A(t))^{-1}\|^3 d\omega = 2\|\dot{A}(t)\|\psi(t).$$

This, (3.8) and (3.9) prove the required result. □

Proof of Theorem 7.3.1: Under (3.3) there is a $t_0 > 0$, such that $x(t) \in \Omega(r)$ for $t \le t_0$. Repeating the reasonings of the proof of Lemma 7.3.4, we get

$$\|x(t)\| \le M_0\|x(0)\| \ \ (t \le t_0).$$

But $M_0\|x(0)\| < r$. So we can extend this inequality to $[0, \infty)$. □

Lemma 7.3.6 *The inequalities $\psi(t) \le \tilde{\psi}(t)$ and $\zeta(t) \le \tilde{\zeta}(t)$ are true.*

Proof: Thanks to Lemma 1.5.7, for any constant $n \times n$-matrix A_0 the inequality

$$\|(I\lambda - A_0)^{-1}\| \le \frac{(N^2(A_0) - 2Re\,(\bar{\lambda}\,Trace\,(A_0)) + n|\lambda|^2)^{(n-1)/2}}{|det\,(\lambda I - A_0)|(n-1)^{(n-1)/2}}$$

is true for any regular λ of A_0. Since $A(t)$ is real and $Im\,Trace\,A(t) = 0$, we have

$$\|(Iiy - A(t))^{-1}\| \le \frac{(N^2(A(t)) + ny^2)^{(n-1)/2}}{|det\,(iyI - A(t))|\ (n-1)^{(n-1)/2}}\ (y \in \mathbf{R}).$$

Hence the required result follows. □

Proof of Corollary 7.3.2: Let condition (3.4) hold. With an $\epsilon > 0$ substitute the equality

$$x(t) = u_\epsilon(t)e^{-\epsilon t}$$

into (1.1). Then we have the equation

$$\dot{u}_\epsilon(t) = (A(t) + I\epsilon)u_\epsilon(t) + F_\epsilon(u_\epsilon(t), t)\ (t \ge 0),$$

where

$$F_\epsilon(u_\epsilon(t), t) = e^{\epsilon t}F(u_\epsilon(t)e^{-\epsilon t}, t).$$

Due to (1.2),

$$\|F_\epsilon(h, t)\| \le \nu(t)\|h\|\ (h \in \Omega(r), t \ge 0).$$

Put

$$\psi_\epsilon(t) = \frac{1}{\pi}\int_{-\infty}^{\infty} \|(A(t) - iI(\omega - \epsilon i))^{-1}\|^3 d\omega$$

and

$$\zeta_\epsilon(t) = \frac{1}{\pi}\int_{-\infty}^{\infty} \|(A(t) - i(\omega - \epsilon i)I)^{-1}\|^2 d\omega\ (t \ge 0).$$

Due to (3.4) we can take ϵ sufficiently small, such that

$$\|\dot{A}(t)\|\psi_\epsilon(t) + \nu(t)\zeta_\epsilon(t) \le 1\ (t \ge 0).$$

Thanks to Theorem 7.3.1, there is a constant M_ϵ, such that

$$\|u_\epsilon(t)\| \le M_\epsilon\|x_0\|.$$

Thus

$$\|x(t)\| \le M_\epsilon\|x_0\|e^{-\epsilon t}\ (t \ge 0).$$

This proves Corollary 7.3.2.

7.3.3 Absolute Stability and Region of Attraction

Definition 7.3.7 *We will say that the zero solution to equation (1.1) is absolutely stable in the class of nonlinearities*

$$\|F(h,t)\| \leq \nu(t)\|h\| \ \ (h \in \mathbf{C}^n, t \geq 0), \tag{3.11}$$

if there exist a M_0 independent of the specific form of the function F (but dependent on $\nu(t)$) such that for any solution $x(t)$ of (1.1) inequality (3.2) is fulfilled.

Theorem 7.3.1 implies

Corollary 7.3.8 *Let condition (3.1) hold. Then the zero solution to equation (1.1) is absolutely stable in the class of nonlinearities (3.11).*

Furthermore,

$$\|(A(t) - iIy)^{-1}h\| \geq \frac{\|h\|}{\|A(t) - iIy\|} \geq$$

$$\frac{\|h\|}{\|A(t)\| + |y|} \ \ (h \in \mathbf{R}^n, y \in \mathbf{R}).$$

Thus, according to (3.6)

$$(W(t)h,h) \geq \frac{1}{\pi}\int_{-\infty}^{\infty} \frac{\|h\|^2 dy}{(\|A(t)\| + |y|)^2} =$$

$$\frac{2\|h\|^2}{\pi\|A(t)\|}.$$

Moreover, according to the definition of $W(t)$, we easily have

$$\|W(t)\| \leq \zeta(t) \ \ (t \geq 0).$$

Hence, due to (3.9)

$$\frac{2\|x(t)\|^2}{\pi\|A(t)\|} \leq \zeta(0)\|x(0)\|^2.$$

We thus get

Lemma 7.3.9 *Under (3.6), let condition (1.2) be fulfilled with $r \leq \infty$. Then with the notations*

$$a_0 := \sup_{t \geq 0} \|A(t)\| \ \text{and} \ M_0 = \sqrt{\frac{a_0\zeta(0)\pi}{2}},$$

inequality (3.2) is valid, provided (3.3) holds.

Due to Lemma 7.3.6, $\zeta(0) \le \tilde{\zeta}(0)$. In addition, as it is shown in Subsection 7.3.1,

$$\tilde{\zeta}(0) \le \zeta_0(0) = \frac{2^{n-1}N^{2(n-1)}(A(0))}{|\alpha(0)|^{2n-1}(n-1)^{n-1}}.$$

Now Corollary 7.3.2 implies

Corollary 7.3.10 *Under conditions (1.2) and (3.4), the zero solution to equation (1.1) is exponentially stable. Moreover, let* $M_0 = M_1$*, where*

$$M_1 = \frac{N^{n-1}(A(0))}{|\alpha(0)|^{n-1/2}}\sqrt{2^{n-2}a_0\pi(n-1)^{-n+1}}. \tag{3.12}$$

Then any vector x_0 *satisfying (3.3), belongs to the region of attraction of the zero solution.*

Example 7.3.11 *Let us consider the system*

$$\begin{aligned}\dot{x}_1 &= -3a(t)x_1 + b(t)x_2 + f_2(x_1, x_2, t),\\ \dot{x}_2 &= a(t)x_2 - b(t)x_1 + f_2(x_1, x_2, t),\end{aligned} \tag{3.13}$$

where $a(t), b(t)$ are positive differentiable functions, $f_k(x_1, x_2, t)$ $(k = 1, 2)$ are continuous functions defined on $\Omega(r) \times [0, \infty)$ with

$$\Omega(r) = \{z \in R^2 : \|z\| \le r\}.$$

In addition,

$$f_1^2(h_1, h_2, t) + f_2^2(h_1, h_2, t) \le \nu^2(t)(h_1^2 + h_2^2)$$
$$(h = (h_1, h_2) \in \Omega(r), t \ge 0)$$

and

$$3a^2(t) < b^2(t) < 4a^2(t).$$

Then

$$\|\dot{A}(t)\| \le 3|\dot{a}(t)| + |\dot{b}(t)|$$

and

$$\alpha(t) = -a(t) + \sqrt{4a^2(t) - b^2(t)},\, N^2(A(t)) = 10a^2(t) + 2b^2(t).$$

According to (3.5),

$$\psi_0(t) = \frac{16(5a^2(t) + b^2(t))^{3/2}}{\pi|\alpha(t)|^5}.$$

and

$$\zeta_0(t) = \frac{2(10a^2(t) + 2b^2(t))}{|\alpha(t)|^3}.$$

Thanks to Corollary 7.3.4, system (3.13) is stable, provided (3.1) holds. Besides, Corollary 7.3.10 gives us a bound for the region of attraction of the zero solution.

7.4 Additional Stability Conditions

In this section as well as in the previous section, we consider system (1.1) with a real differentiable matrix $A(t) = (a_{jk}(t))_{j,k=1}^n$, which is Hurwitzian for each $t \geq 0$ and uniformly bounded on $[0, \infty)$. But stability conditions are formulated in terms of the individual eigenvalues of the variable matrix.

7.4.1 Stability Criteria

Denote

$$\eta(t) := \int_0^\infty \|e^{A(t)s}\|^2 \, ds.$$

Theorem 7.4.1 *Let the conditions (1.2) and*

$$\|\dot{A}(t)\|\eta^2(t) + 2\nu(t)\eta(t) \leq 1 \ \ (t \geq 0) \tag{4.1}$$

hold. Then the zero solution to equation (1.1) is stable. Moreover, there is a constant $M_0 \geq 1$ such that (3.2) is valid for any solution of (1.1), provided (3.3) holds.

The proof of this theorem is presented in the next subsection. Below we suggest additional estimates for M_0. Repeating the arguments of the proof of Corollary 7.3.2 we can show that Theorem 7.4.1 implies the following result.

Corollary 7.4.2 *Let the conditions (1.2) and*

$$\sup_{t \geq 0} \ \|\dot{A}(t)\|\eta^2(t) + 2\nu(t)\eta(t) < 1 \tag{4.2}$$

hold. Then the zero solution to equation (1.1) is exponentially stable. Moreover, there is a constant $M_0 \geq 1$ such that estimate (3.2) is valid, provided condition (3.3) holds.

Recall that $g(A)$ is introduced in Section 1.5 and $\alpha(t)$ is defined in the previous section. Put

$$\tilde{\eta}(t) = \sum_{j,k=0}^{n-1} \frac{g^{j+k}(A(t)) \, (k+j)!}{2^{j+k+1}|\alpha(A((t))|^{j+k+1}(j! \, k!)^{3/2}}.$$

Due to Lemma 1.9.1, $\eta(t) \leq \tilde{\eta}(t)$. Now the previous theorem implies

Corollary 7.4.3 *Let the conditions (1.2) and*

$$\|\dot{A}(t)\|\tilde{\eta}^2(t) + 2\nu(t)\tilde{\eta}(t) \leq 1 \ \ (t \geq 0)$$

hold. Then the zero solution to equation (1.1) is stable. Moreover, there is a constant $M_0 \geq 1$ such that (3.2) is valid for any solution of (1.1), provided (3.3) holds.

In particular, if $n = 2$, then

$$\tilde{\eta}(t) = \frac{1}{2|\alpha(A(t)|}(1 + \kappa(t) + \frac{\kappa^2(t)}{2}). \quad (4.3)$$

Here and below

$$\kappa(t) := \frac{g(A(t))}{|\alpha(A(t))|}.$$

If $n = 3$, then

$$\tilde{\eta}(t) = \frac{1}{2|\alpha(A(t)|}(1 + \kappa + \frac{(\sqrt{2}+1)}{2}\kappa^2 + \frac{3\sqrt{2}\kappa^3}{4} + \frac{2\kappa^4}{3}) \ (\kappa = \kappa(t)). \quad (4.4)$$

7.4.2 Proof of Theorem 7.4.1

Lemma 7.4.4 *Let conditions (1.2) hold with $r = \infty$ and (4.1) hold. Then there is a constant M_0, such that (3.2) holds.*

Proof: Recall that

$$W(t) := 2\int_0^\infty e^{A(t)s}e^{A^*(t)s}ds = \frac{1}{\pi}\int_{-\infty}^\infty (-iI\omega - A_0^*(t))^{-1}(iI\omega - A_0(t))^{-1}d\omega$$

is a solution of equation (3.7) (see Section 1.9). So $2\eta(t) = \zeta(t)$, where $\zeta(t)$ is defined in the previous section. Condition to (3.8) provides inequality (3.9). Moreover, as its is shown in Subsection 7.3.3,

$$(W(t)h, h) \geq \frac{2\|h\|^2}{\pi\|A(t)\|}.$$

Thus (3.9) implies

$$2(x(t), x(t)) \leq \|A(t)\|\pi(W(0)x(0), x(0)). \quad (4.5)$$

As it was shown in the proof of Theorem 3.4.1, $\|\dot{W}(t)\| \leq \|\dot{A}(t)\|\eta^2(t)$ and $\|W(t)\| \leq 2\eta(t) \ \ (t \geq 0)$. Now inequalities (3.8) and (4.5) prove the lemma. □

Lemma 7.4.5 *Under (4.4), let condition (1.2) be fulfilled with $r \leq \infty$. Then there is a constant M_0, such that (3.2) is valid, provided (3.3) holds.*

Proof: Under (3.3), $\|x(0)\| < r$ and there is a $t_0 > 0$, such that $x(t) \in \Omega(r)$ for $t \leq t_0$. Repeating the arguments of the proof of Lemma 7.4.4, we get

$$\|x(t)\| \leq M_0\|x(0)\| \ \ (t \leq t_0).$$

But $M_0\|x(0)\| < r$. So we can extend this inequality to $[0, \infty)$. □

The assertion of Theorem 7.4.1 follows from Lemmas 7.4.4 and 7.4.5.

7.4.3 Absolute Stability and Region of Attraction

Theorem 7.4.1 implies

Corollary 7.4.6 *Let condition (4.1) hold. Then the zero solution to equation (1.1) is absolutely stable in the class of nonlinearities (3.11).*

Due to (4.5)

$$\frac{\|x(t)\|^2}{\pi\|A(t)\|} \leq \eta(0)\|x(0)\|^2.$$

Now Lemma 7.4.4 implies

Lemma 7.4.7 *Under (2.1), let condition (1.2) be fulfilled with $r \leq \infty$. Then with*

$$a_0 := \sup_{t\geq 0} \|A(t)\| \text{ and } M_0 = \sqrt{a_0\eta(0)\pi}$$

inequality (3.2) is valid, provided (3.3) holds.

Since $\eta(0) \leq \tilde{\eta}(0)$, Corollary 7.4.2 implies

Corollary 7.4.8 *Under conditions (1.2) and (4.3) the zero solution to equation (1.1) is exponentially stable. Moreover, let $M_0 = \tilde{M}_0$, where*

$$\tilde{M}_0 = \sqrt{\tilde{\eta}(0)\pi a_0}.$$

Then the vectors $x(0)$ satisfying (3.3), belongs to the region of attraction.

7.4.4 Examples

Example 7.4.9 *Let $A(t) = (a_{jk}(t))_{j,k=1}^2$ be a real differentiable 2×2 matrix.*

In the general case, according to Example 1.6.1, $g(A(t)) \leq |a_{12}(t) - a_{21}(t)|$. If the eigenvalues of $A(t)$ are nonreal, then as it is shown in that example,

$$g(A(t)) = \sqrt{(a_{11}(t) - a_{22}(t))^2 + (a_{21}(t) + a_{12}(t))^2}. \tag{4.6}$$

For instance, let us consider the system

$$\dot{x}_1 = -3a(t)x_1 + x_2 + f_1(x_1, x_2, t),\ \dot{x}_2 = a(t)x_2 - x_1 + f_2(x_1, x_2, t), \tag{4.7}$$

where $a(t)$ is a positive differentiable function, $f_k(x_1, x_2, t)$ are continuous functions defined on $\Omega(r) \times [0, \infty)$ with $\Omega(r) = \{z \in R^2 : \|z\| \leq r\}$. In addition,

$$f_1^2(h_1, h_2, t) + f_2^2(h_1, h_2, t) \leq \nu^2(t)(h_1^2 + h_2^2)\ \ (h = (h_1, h_2) \in \Omega_r, t \geq 0)$$

and

$$a(t) \leq 1/2\ \ (t \geq 0).$$

Then $\|\dot{A}(t)\| = 3|\dot{a}(t)|$, $\alpha(A(t)) = -a(t)$ and due to (4.1) $g(A(t)) = 4a(t)$. Thus (4.6) implies

$$\tilde{\eta}(t) = \frac{13}{2a(t)}.$$

Thanks to Corollary 7.4.2 system (4.7) is stable, provided

$$\frac{3 \cdot 13^2 |\dot{a}(t)|}{4a^2(t)} + \frac{13\nu(t)}{a(t)} \le 1 \ \ (t \ge 0).$$

Besides, Corollary 7.4.8, gives us a bound for the region of attraction.

Example 7.4.10 *Let* $A(t) = (a_{jk}(t))_{j,k=1}^3$ *be a real differentiable* 3×3-*matrix.*

Consider system (1.1) with condition (1.2) and $n = 3$. According to inequality (5.3) from Section 1.5,

$$g(A(t)) \le v(t) := [(a_{12} - a_{21})^2 + (a_{13} - a_{31})^2 + (a_{23} - a_{32})^2]^{1/2}.$$

For simplicity assume that

$$a_{jj}(t) + \sum_{k=1, k \ne j}^{3} |a_{jk}(t)| \le -\rho_0(t) \ \ (t \ge 0, j = 1, 2, 3)$$

where ρ_0 is a positive scalar function. Then due to the well known result from the book (Marcus and Minc, 1964, Section 3.3.5), $\alpha(A(t)) \le -\rho_0(t)$.

Thus (4.4) implies

$$\tilde{\eta}(t) \le \eta_1(t) := \frac{1}{\rho_0}(1 + \kappa_1 + \frac{(\sqrt{2}+1)\kappa_1^2}{2} + \frac{3\sqrt{2}\kappa_1^3}{4} + \frac{2\kappa_1^4}{3}),$$

where $\kappa_1(t) = \frac{v(t)}{\rho_0(t)}$. Thanks to Corollary 7.4.3 system (1.1) under consideration is stable, provided

$$\eta_1^2 \|\dot{A}(t)\| + 2\eta_1(t)\nu(t) \le 1 \ \ (t \ge 0).$$

In addition, Corollary 7.4.8, gives us a bound for the region of attraction.

8. Essentially Nonlinear Systems

8.1 The Freezing Method for Nonlinear Systems

8.1.1 Stability Conditions

As above, $\|.\|$ is the Euclidean norm and $\Omega\ (r)$ is the ball in $\mathbf{C}^n$ with a radius $r \leq \infty$ and with the center at zero. Let

$$B(h,t) = (b_{jk}(h,t))_{j,k=1}^n$$

be an $n \times n$-matrix for every $h \in \Omega(r)$ and $t \geq 0$. Everywhere below it is assumed that $B(h,t)$ *continuously depends on* $h \in \Omega(r)$ and $t \geq 0$.

Let us consider in $\mathbf{C}^n$ the equation

$$\dot{x}(t) = B(x(t),t)x(t) \ \ (t \geq 0). \tag{1.1}$$

It is assumed that there are constants $\mu(r), \nu(r)$, such that

$$\|B(h_1,t) - B(h,t)\| \leq \nu(r)\|h_1 - h\| \text{ and}$$

$$\|B(h_1,t) - B(h,s)\| \leq \mu(r)|t-s| \ \ (h, h_1 \in \Omega(r); t, s \geq 0). \tag{1.2}$$

Put

$$M(r) := \sup_{h \in \Omega(r), t \geq 0} \|B(h,t)h\| \text{ and } q_0(r) = \nu(r)M(r) + \mu(r).$$

Assume that

$$p(r,t) := \sup_{s \geq 0, h \in \Omega(r)} \|exp\,[B(h,s)]t]\| < \infty \tag{1.3}$$

is a continuous function of t, such that

$$\chi(r) := \sup_{t \geq 0} p(r,t) < \infty. \tag{1.4}$$

Theorem 8.1.1 *For a positive* $r < \infty$, *let the conditions (1.2) - (1.4) and*

$$\theta(r) := q_0(r) \int_0^\infty tp(r,t)dt < 1 \tag{1.5}$$

hold. Then the zero solution to equation (1.1) is stable. Moreover, any solution $x(t)$ of (1.1) is subject to the estimate

$$\|x(t)\| \leq \|x(0)\| \frac{\chi(r)}{1-\theta(r)} \text{ for all } t \geq 0, \tag{1.6}$$

provided

$$\|x(0)\| \frac{\chi(r)}{1-\theta(r)} < r. \tag{1.7}$$

The proofs of this theorem and the next one are presented in Section 8.2.

Recall that the quantities $g(A)$ and $\alpha(A)$ are introduced in Sections 1.5 and 1.3, respectively. Furthermore, let us suppose that

$$\rho(r) := - \sup_{h \in \Omega(r), t \geq 0} \alpha(B(h,t)) > 0 \tag{1.8}$$

and

$$v(r) := \sup_{h \in \Omega(r), t \geq 0} g(B(h,t)) < \infty. \tag{1.9}$$

Put

$$\tilde{\chi}(r) := \sup_{t \geq 0} e^{-\rho(r)t} \sum_{k=0}^{n-1} \frac{v^k(r) t^k}{(k!)^{3/2}}.$$

Due to Corollary 1.5.3 $\chi(r) \leq \tilde{\chi}(r)$ and

$$\int_0^\infty t p(r,t) dt \leq \sum_{j=0}^{n-1} \frac{(j+1) v^j(r)}{\sqrt{j!} \rho^{j+2}(r)}.$$

Now Theorem 8.1.1 and a small perturbation imply

Corollary 8.1.2 *For a positive $r < \infty$, let the conditions (1.2), (1.8), (1.9) and*

$$\Gamma(r) := q_0(r) \sum_{j=0}^{n-1} \frac{(j+1) v^j(r)}{\sqrt{j!} \rho^{j+2}(r)} < 1 \tag{1.10}$$

hold. Then the zero solution to equation (1.1) is exponentially stable. Moreover, with $\theta(r) = \Gamma(r)$ and $\chi(r) = \tilde{\chi}(r)$, any solution $x(t)$ of (1.1) is subject to estimate (1.6), provided (1.7) holds.

Example 8.1.3 *Let us consider the system*

$$\dot{x} = -\cos\,(ax)x - 2y - \sin\,(ax)y,$$
$$\dot{y} = -\cos\,(ay)y + 2x + \sin\,(ay)x \quad (0 < a = const < 1). \tag{1.11}$$

Rewrite this system in the form (1.1) with $B(h,t) = B(h) = (b_{jk})_{j,k=1}^2$ $(b_{jk} = b_{jk}(h))$, where $h = (x,y)$,

$$b_{11} = -\cos(ax), b_{12} = -2 - \sin(ax), b_{21} = 2 + \sin(ay),\ b_{22} = -\cos(ay).$$

We have $\lambda_{1,2}(B(h)) = T \pm (T^2 - D)^{1/2}$, where

$$T = T(h) = (b_{11} + b_{22})/2 = -(\cos(ay) + \cos(ax))/2$$

and

$$D = D(h) = b_{11}b_{22} - b_{12}b_{21} = \cos(ay)\cos(ax) + (2 + \sin(ax))(2 + \sin(ay)).$$

Take $r = 1$. Then $\cos(ax) \geq \cos a \geq \cos \pi/3 = 1/2$ $(|x| < 1)$ and

$$T^2 - D \leq (\cos(ay) + \cos(ax))^2/4 - \cos(ay)\cos(ax) - 1 =$$

$$(\cos(ay) - \cos(ax))^2/4 - 1 \leq 0.$$

So

$$Re\ \lambda_{1,2}(B(h)) = -(\cos(ay) + \cos(ax))/2 \leq -\rho = -1/2\ (|x|, |y| \leq 1).$$

Due to Example 1.6.1,

$$g(B(h)) \leq |b_{12} - b_{21}| = |4 + \sin(ay) + \sin(ay)| \leq v = 6$$

$(|x|, |y| \leq 1)$. Moreover,

$$\|B(h) - B(h_1)\|^2 \leq \sum_{j,k=1}^{n} (b_{jk}(h) - b_{jk}(h_1)^2 \leq \nu^2(r) = 16a^2\ \ (|x|, |y| \leq 1).$$

In addition,

$$M(r) = M(1) \leq \sup_{h \in \Omega(1)} N(B(h)) \leq (2 + (2 + \sin(a))^2)^{1/2} \leq \sqrt{11}.$$

Thus,

$$\Gamma(r) \leq \nu(r)M(r)(1/\rho^{-2} + 2v\rho^{-3}) \leq 4a\sqrt{11}(4 + 96).$$

For a sufficiently small a we have stability condition (1.10).

8.1.2 Lyapunov's Exponents

Denote by $z_0(r)$ the extreme right-hand root of the algebraic equation

$$z^{n+1} = q_0(r) \sum_{j=0}^{n-1} \frac{v^j(r)(j+1)}{\sqrt{j!}} z^{n-j-1}. \tag{1.12}$$

Theorem 8.1.4 *For some positive $r \leq \infty$, let conditions (1.2) and (1.9) hold. If, in addition, the matrix $B(h,t) + I(z_0(r) + \epsilon)$ is Hurwitzian for an $\epsilon > 0$, all $h \in \Omega(r)$ and $t \geq 0$, then the zero solution of (1.1) is exponentially stable. Besides, the relations (1.8) and (1.10) hold. Moreover, with $\theta(r) = \Gamma(r)$ and $\chi(r) = \tilde{\chi}(r)$ any initial vector satisfying condition (1.7), belongs to the region of attraction of the zero solution and the corresponding solution of (1.1) is subject to the estimates (1.6) and*

$$\|x(t)\| \leq const\ exp\,[(-\rho(r)t + z_0(r))t] \quad (t \geq 0).$$

Let $v(r) \neq 0$. Put

$$w_n = \sum_{k=0}^{n-1} \frac{k+1}{\sqrt{k!}}.$$

Setting $z = v(r)y$ in (1.12) and applying Lemma 1.11.1, we can assert that $z_0(r) \leq \delta_0(r)$, where

$$\delta_0(r) = \begin{cases} v^{1-2/(n+1)}(r)[q_0(r)w_n]^{1/(n+1)} & \text{if } q_0(r)w_n \leq v^2(r), \\ \frac{q_0(r)w_n}{v(r)} & \text{if } q_0(r)w_n > v^2(r) \end{cases}$$

Thus in the previous theorem we can replace $z_0(r)$ by $\delta_0(r)$.

8.2 Proofs of Theorems 8.1.1 and 8.1.4

Proof of Theorem 8.1.1: Let us introduce the linear equation

$$dy_h/dt = B(h(t),t)y_h, \tag{2.1}$$

where $h(t) : [0,\infty) \to \Omega(r)$ is a differentiable function. If $h(t) = x(t)$ is a solution of (1.1), then (2.1) and (1.1) coincide.

Further, the continuous dependence of solutions on initial data implies that under the condition $\|x(0)\| < r$, there is t_0, such that

$$x(t) \in \Omega(r) \text{ for } 0 \leq t \leq t_0 \tag{2.2}$$

for a solution $x(t)$ of (1.1). If we put $A(t) = B(h(t),t)$ with some function $h(t)$, then (2.1) takes the form $\dot{x} = A(t)x$. Due to (1.2) we have

$$\|B(h(t),t) - B(h(s),s)\| \leq \|B(h(t),t) - B(h(s),t)\| +$$
$$\|B(h(s),t) - B(h(s),s)\| \leq$$
$$\nu(r)\|h(t) - h(s)\| + \mu(r)|t-s| \quad (t,s \leq t_0).$$

Hence,

$$\|B(x(t),t) - B(x(s),s)\| \leq \nu(r)\|x(t) - x(s)\| + \mu(r)|t-s| \leq$$

$$(\nu(r) \sup_{t \le t_0} \|\dot{x}(t)\| + \mu(r))|t - s| \ (s, t \le t_0).$$

But according to (1.1) $\|\dot{x}(t)\| \le M(r)$ $(t \le t_0)$. Thus,

$$\|B(x(t), t) - B(x(s), s)\| \le q_0(r)|t - s| \ (s, t \le t_0).$$

Applying Corollary 3.1.2 with $A(t) = B(x(t), t)$ to equation (1.1) we have the bound

$$\|x(t)\| \le \chi(r)(1 - \theta(r))^{-1}\|x(0)\| \text{ for } t \le t_0.$$

But condition (1.7) allows us to extend this bound to all $t \ge 0$. So estimate (1.6) is proved. It yields the Lyapunov stability.

Proof of Theorem 8.1.4: Repeating the reasoning of the proof of Theorem 3.1.5, one can show that the Hurwitzness of the matrix $B(h, t) + z(r)I$ implies inequality (1.5). Now Corollary 8.1.2 yields the required result. □

8.3 Perturbations of Nonlinear Systems

Consider in $\mathbf{C}^n$ the equation

$$\dot{x}(t) = B(x(t), t)x(t) + C(x(t), t)x(t) \ \ (t \ge 0), \tag{3.1}$$

where $B(z, t)$ and $C(z, t)$ are $n \times n$-matrices continuous in $z \in \mathbf{C}^n$ and $t \ge 0$.

Let $U_h(t, s)$ $(t, s \ge 0)$ be the evolution operator of the linear equation

$$\dot{y}(t) = B(h(t), t)y(t) \ \ (t \ge 0) \tag{3.2}$$

with a differentiable function $h : \ [0, \infty) \to \Omega(r)$. Let us assume that there are positive continuous functions $\phi(r, t, s)$ and $m(t, r)$ independent of h, such that

$$\|U_h(t, s)\| \le \phi(r, t, s) \ \text{ and } \|C(w, t)\| \le m(t, r)$$
$$(w \in \Omega(r); \ t, s \ge 0). \tag{3.3}$$

Additionally, suppose that

$$\varpi(r) := \sup_{t \ge 0} \phi(r, \ t, 0) < \infty \text{ and}$$

$$\eta(r) \equiv \sup_{t \ge 0} \int_0^t \phi(r, t, s)m(s, r)ds < 1. \tag{3.4}$$

Theorem 8.3.1 *Let conditions (3.3) and (3.4) hold. Then the zero solution of equation (3.1) is stable. If, in addition, an initial vector $x(0)$ satisfies the inequality*

$$\varpi(r)\|x(0)\| < r(1 - \eta(r)), \tag{3.5}$$

then the corresponding solution $x(t)$ of (3.1) is subject to the estimate

$$\|x(t)\| \le \frac{\|x(0)\|\varpi(r)}{1 - \eta(r)} \ \ (t \ge 0).$$

Proof: By the property of the evolution operator, equation (3.2) is equivalent to the following one:

$$y(t) = U_h(t,0)y(0) + \int_0^t U_h(t,s)C(h(s),s)y(s)ds \ \ (t \geq 0).$$

Taking $h(t) = x(t)$-the solution of (3.1), we get

$$x(t) = U_x(t,0)x(0) + \int_0^t U_x(t,s)C(x(s),s)x(s)ds \ \ (t \geq 0).$$

Clearly, (3.5) implies $\|x(0)\| < r$. Then for a sufficiently small $t_0 > 0$ we have

$$\|x(t)\| \leq r \quad (t \leq t_0).$$

So

$$\|x(t)\| \leq \|U_x(t,0)\|\|x(0)\| + \int_0^t \|U_x(t,s)\|m(s,r)\|x(s)\|ds \leq$$

$$\varpi(r)\|x(0)\| + \int_0^t \phi(r,t,s)m(s,r)\|x(s)\|ds \ \ (t \leq t_0).$$

Hence,

$$\sup_{0 \leq t \leq t_0} \|x(t)\| \leq \varpi(r)\|x(0)\| + \sup_{0 \leq t \leq t_0} \|x(t)\|\eta(r),$$

and therefore,

$$\sup_{0 \leq t \leq t_0} \|x(t)\| \leq \frac{\varpi(r)\|x(0)\|}{1-\eta(r)} \leq r.$$

According to (3.5) this inequality can be extended to $[0,\infty)$. This proves the result. □

8.4 Nonlinear Triangular Systems

Let us consider the triangular system

$$\dot{u}_j(t) = \sum_{k=1}^{j} a_{jk}(u,t)u_k(t)$$

$$(u = (u_k(t))_{k=1}^n, \ t \geq 0; \ j = 1, ..., n) \tag{4.1}$$

with continuous real functions

$$a_{jk}(.,.) : \mathbf{R}^n \times [0,\infty) \to \mathbf{R} \quad (1 \leq k \leq j \leq n).$$

For a positive $r \leq \infty$ put

$$\alpha(t,r) := \max_{w\in\Omega(r)} \max_{j=1,\dots,n} a_{jj}(w,t) \tag{4.2}$$

$$g(t,r) := \max_{w\in\Omega(r)} \sqrt{\sum_{1\le k<j\le n} a_{jk}^2(w,t)} \tag{4.3}$$

and

$$\psi_r(t,\tau) := e^{\int_\tau^t \alpha(s,r)ds} \sum_{k=0}^{n-1} \frac{1}{k!}[\int_\tau^t g(t_1,r)dt_1]^k$$

$$(0 \le \tau \le t < \infty),$$

Theorem 8.4.1 *Let the condition*

$$\tilde{\chi}(r) := \sup_{t\ge\tau\ge 0} \psi_r(t,\tau) < \infty \tag{4.4}$$

be fulfilled. Then the zero solution of equation (4.1) is stable. If an initial vector u_0 satisfies the inequality

$$\tilde{\chi}(r)\|u_0\| < r, \tag{4.5}$$

then a solution $u(t)$ of (4.1) with the initial condition $u(\tau) = u_0$ is subject to the estimate

$$\|u(t)\| \le \psi_r(t,\tau)\|u(\tau)\| \quad (t \ge \tau \ge 0). \tag{4.6}$$

Proof: Consider the triangular linear system

$$\dot{v}_j(t) = \sum_{k=1}^{j} a_{jk}(h(t),t)v_k(t), \ (t \ge 0;\ j = 1,\dots,n) \tag{4.7}$$

with a continuous function $h : [0,\infty) \to \Omega(r)$.

By Theorem 2.4.1, the inequality

$$\|v(t)\| \le \psi_r(t,\tau)\|v(\tau)\| \quad (t \ge \tau \ge 0) \tag{4.8}$$

is true for any solution $v(t)$ of system (4.7).

Clearly, $\tilde{\chi}(r) \ge 1$. So condition (4.5) implies the inequality $\|u(\tau)\| < r$. Then for a sufficiently small $t_0 - \tau$,

$$\|u(t)\| \le r \ (0 \le \tau \le t \le t_0).$$

But equation (4.7) coincides with (4.1) if $h(t) = u(t)$, where $u(t)$ is the solution of (4.1). So by (4.8),

$$\|u(t)\| \le \tilde{\chi}(r)\|u_0\| \le r \ (\tau \le t \le t_0).$$

According to (4.6) this inequality can be extended to $[0,\infty)$. This proves the result. □

Corollary 8.4.2 *Let the condition*

$$g(t,r) \leq g_0 < \infty \ (g_0 = const, \ t \geq 0) \tag{4.9}$$

be fulfilled. Then the zero solution to system (4.1) is stable, provided

$$a_{jj}(w,t) \leq \alpha_0 < 0 \ (\alpha_0 = const; \ w \in \Omega(r); \ t \geq 0; \ j = 1, ..., n)$$

8.5 Perturbations of Triangular Systems

Consider system (1.1). It is assumed that $B(w,t) = (b_{jk}(w,t))_{j,k}^n$ is continuous in $w \in \mathbf{R}^n$ and piecewise continuous in t. Also consider the corresponding lower-triangular matrix $A(w,t) = (a_{jk}(w,t))_{j,k=1}^n$ with the entries

$$a_{jk}(w,t) = b_{jk}(w,t) \text{ if } j \geq k; \text{ and } a_{jk}(w,t) = 0 \text{ if } j < k \ \ (w \in \mathbf{C}^n). \tag{5.1}$$

Assume that conditions (4.2-4.4) hold. According to (4.8), the evolution operator of the system (4.7) with a continuous function h defined on $[0,\infty)$ with values in $\Omega(r)$, satisfies the inequality

$$\|U_h(t,s)\| \leq \psi_r(t,s) \ (t,s \geq 0).$$

Put

$$m(t,r) = \sup_{h \in \Omega(r)} \|C(h,t)\|,$$

where $C(h,t) = A(h,t) - B(h,t)$ is the upper triangular matrix with the entries

$$c_{jk}(h,t) = b_{jk}(h,t) \ (j < k) \text{ and } c_{jk}(h,t) = 0 \ (j \geq k).$$

In addition, suppose that

$$\tilde{\eta}(r) \equiv \sup_{t \geq 0} \int_0^t \psi_r(t,s) m(s,r) ds < 1. \tag{5.2}$$

Employing Lemma 8.3.1 and Theorem 8.4.1, we arrive at the following result.

Theorem 8.5.1 *Let condition (5.2) hold. Then the zero solution of equation (1.1) is stable. In addition, if an initial vector u_0 satisfies the inequality*

$$\chi(r)\|u_0\| < r(1 - \tilde{\eta}(r)), \tag{5.3}$$

then the corresponding solution $u(t)$ of system (1.1) is subject to the inequality

$$\|u(t)\| \leq \frac{\chi(r)\|u_0\|}{1 - \tilde{\eta}(r)} \leq r \ (t \geq 0).$$

Note that the following relation holds:

$$m(t,r) \leq \max_{j=1,\dots,n} \sqrt{\sum_{k=j+1}^{n} \sup_{h \in \Omega(r)} |b_{jk}(h,t)|^2}.$$

8.6 Nonlinear Dissipative Systems

Let $\|.\|_{C^n}$ be an arbitrary norm in $\mathbf{C}^n$.

Theorem 8.6.1 *For any $h \in \mathbf{C}^n$ with $\|h\|_{C^n} < r$ and all sufficiently small $\delta > 0$, let*

$$\|I + \delta B(h,t)\|_{C^n} \leq 1 + a(t)\delta \ (t \geq 0) \tag{6.1}$$

where $a(t)$ is a Riemann-integrable function independent of h and having the property

$$\int_0^t a(s)ds \leq 0 \ (t \geq 0). \tag{6.2}$$

Then any solution $x(t)$ of (1.1) satisfies the estimate

$$\|x(t)\|_{C^n} \leq \|x(s)\|_{C^n} \, exp[\int_s^t a(\tau)d\tau] \ \ (t \geq s \geq 0), \tag{6.3}$$

provided $\|x(s)\|_{C^n} < r$.

Proof: The continuous dependence of solutions on initial data implies that under the condition $\|x(0)\| < r$ there is t_0 such that relation (2.2) holds. Again consider equation (2.1) with an arbitrary function $h(t) : [0, \infty) \to \Omega(r)$. Using Theorem 4.1.1 with $A(t) = B(h(t), t)$, we have the required estimate (6.3) for $t \leq t_0$. But condition (6.2) permits us to extend it to the whole positive half-line. □

Let $(.,.)$ be the scalar product in $\mathbf{C}^n$. As above,

$$\Omega(r) = \{h \in \mathbf{C}^n : \|h\| \leq r\},$$

where $\|.\|$ is the Euclidean norm.

Theorem 8.6.2 *For any $t \geq 0$ and $h \in \Omega(r)$, let*

$$Re \ (B(h,t)h, h) \leq \Lambda(t)(h,h), \tag{6.4}$$

where $\Lambda(t)$ is the Riemann-integrable function independent of h. Then under the condition

$$\int_0^t \Lambda(s)ds \leq 0 \ (t \geq 0)$$

any solution $x(t)$ of equation (3.1) satisfies the estimate

$$\|x(t)\| \leq \|x(s)\| exp[\int_s^t \Lambda(\tau) \, d\tau] \ (t > s \geq 0)$$

provided $\|x(s)\| < r$.

Proof: Doing the scalar product in (1.1) by x and considering that

$$\frac{d}{dt}(x(t), x(t)) = 2\, Re(\dot{x}(t), x(t)),$$

we get

$$\frac{d}{dt}(x(t), x(t)) = 2\, Re(B(x(t), t)x(t), x(t)). \tag{6.5}$$

For a sufficiently small t_0 we have relation (2.2) and, therefore,

$$\frac{d}{dt}(x(t), x(t)) \le 2\Lambda(t)(x(t), x(t)),\ 0 \le t \le t_0.$$

Solving this inequality, we arrive at the inequality

$$(x(t), x(t)) \le exp[2\int_0^t \Lambda(s)ds](x(0), x(0)),\ 0 \le t \le t_0.$$

Hence the required result easily follows. □

8.7 Nonlinear Systems with Linear Majorants

Consider in $\mathbf{C}^n$ equation (1.1), where $B(h,t)$ is an $n \times n$-matrix for each $h \in \mathbf{C}^n$ and all $t \ge 0$, again.

For all sufficiently small positive δ and $h \in \Omega(r)$ $(r \le \infty)$, let there be a variable matrix $M(t)$ independent of h, such that the relation

$$|v + \delta B(h,t)v| \le (I + M(t)\delta)|v| \ \ (v \in \mathbf{C}^n,\ t \ge 0) \tag{7.1}$$

is valid. Then we will say that system (1.1) has in set $\Omega(r)$ *the linear majorant* $M(t)$.

Inequality (7.1) means that

$$|b_{jk}(h,t)| \le m_{jk}(t) \text{ for } j \ne k,$$

and

$$|1 + \delta b_{kk}(h,t)| \le 1 + \delta m_{kk}(t)\ (h \in \Omega(r), t \ge 0),$$

and $M(t) = \ (m_{jk}(t))_{j,k=1}^n$. Let us introduce the equation

$$\dot{z}(t) = M(t)z(t)\ (t \ge 0) \tag{7.2}$$

and assume that this equation is stable. This implies that the Cauchy operator $V(t)$ of equation (7.2) is bounded:

$$l = \sup_{t \ge 0} \|V(t)\| < \infty. \tag{7.3}$$

Lemma 8.7.1 *Let system (1.1) have a linear majorant $M(t)$ in the ball $\Omega(r)$. Then, under condition (7.3), any solution $x(t)$ of (1.1) is subject to the inequality*

$$|x(t)| \leq V(t)|x(0)| \; (t \geq 0), \tag{7.4}$$

provided that

$$\|x(0)\| < rl^{-1}. \tag{7.5}$$

Proof: Clearly, $l \geq 1$. So $\|x(0)\| < r$, and condition (2.2) holds. Set $A(t) = B(x(t), t)$, where $x(t)$ is the solution of equation (1.1). Now the inequality (7.4) for $0 \leq t \leq t_0$ is due to Lemma 4.2.1. Since $x(0)$ is in the interior of $\Omega(rl^{-1})$, the solution remains in $\Omega(r)$. This proves the result. □

Corollary 8.7.2 *Let system (1.1) have a linear constant majorant M in the set $\Omega(r)$. In addition, let M be a Hurwitz matrix, such that*

$$l \equiv \sup_{t \geq 0} \|e^{Mt}\| < \infty.$$

Then any solution $x(t)$ of (1.1) with an initial vector $x(0)$ satisfying (7.5) is subject to the inequality

$$|x(t)| \leq exp\,[Mt]|x(0)| \;\; (t \geq 0),$$

and, therefore, the zero solution of equation (1.1) is exponentially stable.

Note that Corollary 1.5.3 yields the inequality

$$l \leq \max_{t \geq 0} e^{\alpha(M)t} \sum_{k=0}^{n-1} \frac{g^k(M) t^k}{(k!)^{3/2}}.$$

8.8 Second Order Vector Systems

Again let $\mathbf{R}^n$ $(n > 1)$ be a real Euclidean space with the scalar product $(.,.)$, the unit matrix I and the norm $\|.\| = \sqrt{(.,.)}$. For a positive $r \leq \infty$ put

$$\omega_r := \left\{ v, h \in R^n : \|v\|^2 + \|h\|^2 \leq r^2 \right\}$$

and consider the equation

$$\ddot{x} + 2F(x, \dot{x}, t)\dot{x} + G(x, \dot{x}, t)x = 0 \tag{8.1}$$

with real $n \times n$-matrices $F(h, w, t)$ and $G(h, w, t)$ continuously dependent on $(h, w) \in \omega_r$ and $t \geq 0$. Take the initial conditions

$$x(t) = x_0, \dot{x}(0) = x_1 \;\; (x_0, x_1 \in \mathbf{R}^n). \tag{8.2}$$

A solution of the problem (8.1), (8.2) is a twice continuously differentiable function $x : [0,\infty) \to \mathbf{R}^n$, satisfying (8.1) and (8.2) for all $t \geq 0$. Under consideration, problem (8.1), (8.2) has solutions, due to the solution estimates established below, since the entries continuously depend on the arguments.

Recall that

$$A_R = (A + A^*)/2 \text{ and } A_I = (A - A^*)/2i$$

are *the real and imaginary components* of a matrix A, respectively and A^* is the matrix adjoint to A. Let A_0 be a Hermitian matrix. Then we will write $A_0 \geq 0$ ($A_0 > 0$) if it is positive definite (strongly positive definite). For two Hermitian matrices A_0, B_0 we write $A_0 \geq B_0$ if $A_0 - B_0 \geq 0$, and $A_0 > B_0$ if $A_0 - B_0 > 0$.

So

$$F_R(h,w,t) \equiv (F(h,w,t) + F^*(h,w,t))/2,$$

$$G_R(h,w,t) \equiv (G(h,w,t) + G^*(h,w,t))/2.$$

It is assumed that there is a constant $m_F(r) > 0$, such that

$$m_F(r)I \leq F_R(h,w,t) \text{ and } 0 \leq G_R(h,w,t) \leq m_F(r)(2F_R(h,w,t) - m_F(r)I)$$

$$(\,(h,w) \in \omega_r;\ t \geq 0). \tag{8.3}$$

Put

$$\Psi(h,w,t) := 2m_F(r)F(h,w,t) - G(h,w,t). \tag{8.4}$$

According to the above definitions we write

$$\Psi_R(h,w,t) = (\Psi + \Psi^*)/2,\ \Psi_I(h,w,t) = (\Psi - \Psi^*)/2i$$

with $\Psi = \Psi(h,w,t)$.

Theorem 8.8.1 *Under condition (8.3), let*

$$\zeta_0(r) := \sup_{(h,w)\in\omega_r,\ t\geq 0} \|\Psi_R(h,w,t)\| + \|\Psi_I(h,w,t)\| < 2m_F^2(r). \tag{8.5}$$

Then the zero solution to equation (8.1) is exponentially stable. Moreover, the initial vectors satisfying the condition

$$\|x_0\|^2 + \|x_1\|^2 < r^2\ max\ \{\zeta_0(r) - m_F^2(r),\ \frac{1}{\zeta_0(r) - m_F^2(r)}\}, \tag{8.6}$$

belong to the region of attraction of the zero solution and the corresponding solution $x(t)$ of problem (8.1), (8.2) satisfies the inequality

$$(\zeta_0(r) - m_F^2(r))\|x(t)\|^2 + \|\dot{x}(t)\|^2 \leq$$

$$e^{2\Lambda_0(r)t}((\zeta_0(r) - m_F^2(r))\|x_0\|^2 + \|x_1\|^2)\ \ (t \geq 0) \tag{8.7}$$

with $\Lambda_0(r) := -m_F(r) + \sqrt{\zeta_0(r) - m_F^2(r)} < 0$.

This theorem is proved in the next section.

Corollary 8.8.2 *For some positive r and all $(h,w) \in \omega_r$, and $t \geq 0$, let matrices $F(h,w,t)$ and $G(h,w,t)$ be symmetric, and there be a constant $m_F(r) > 0$, such that*

$$m_F(r)I \leq F(h,w,t) \text{ and } m_F^2(r)I \leq 2m_F(r)F(h,w,t) - G(h,w,t) <$$

$$2m_F^2(r)I \ \ ((h,w) \in \omega_r;\ t \geq 0).$$

Then the zero solution to equation (8.1) is exponentially stable. Moreover, the initial vectors satisfying condition (8.6) with

$$\zeta_0(r) = \sup_{(h,w)\in\omega_r, t\geq 0} \|2m_F(r)F(h,w,t) - G(h,w,t)\|$$

belong to the region of attraction of the zero solution and the corresponding solution $x(t)$ of problem (8.1), (8.2) satisfies the inequality (8.7).

Example 8.8.3 *Consider the equation*

$$\ddot{x} + 2f(x,\dot{x},t)\dot{x} + g(x,\dot{x},t)x = 0 \tag{8.8}$$

with positive continuous scalar functions f and g defined on $\omega_{2,r} \times [0,\infty)$ with

$$\omega_{2,r} := \{w, s \in R^1 : w^2 + s^2 \leq r^2\}.$$

Let

$$m_f(r) := \inf_{(s,w)\in\omega_{2,r}, t\geq 0} f(s,w,t) > 0.$$

Then due to Corollary 8.8.2 the zero solution to equation (8.8) is exponentially stable, provided

$$m_f^2(r) \leq 2m_f(r)\ f(s,w,t) - g(s,w,t) \leq \tilde{p}(r) < 2m_f^2(r)$$

$$(\tilde{p}(r) = const;\ (s,w) \in \omega_{2,r}, t \geq 0).$$

In particular, take $r = 1$,

$$f(s,w,t) = 2 + \frac{w^2}{2} sin^2\ (s+t) \text{ and } g(s,w,t) = 3 + \frac{w^3}{2} cos\ (s^2 + t^2).$$

Then $m_f(1) = 2$ and

$$m_f^2(1) = 4 \leq 2m_f(1)\ f(s,w,t) - g(s,w,t) = 8 + 2w^2\ sin^2\ (s+t) - 3-$$

$$\frac{w^3}{2} cos\ (t^2 + s^2) < 2m_f^2(1) = 8 \ \ (t \geq 0;\ |w|, |s| < 1).$$

So the zero solution to equation (8.8) is exponentially stable.

Example 8.8.4 *Let $F(h,w,t) = (f_{jk})_{j,k=1}^n$ $(f_{jk} \equiv f_{jk}(h,w,t))$ be a real symmetric matrix, having the properties*

$$f_{jj}(h,w,t) \geq 9/8, \sum_{k=1,k\neq j}^{n} |f_{jk}(h,w,t)| \leq 1/8 \ \ (j=1,...,n). \tag{8.9}$$

Here and below in this section $t \geq 0$ and $(h,w) \in \omega_r$ for some positive $r \leq \infty$. Then

$$\inf_j f_{jj}(h,w,t) - \sum_{k=1,k\neq j}^{n} |f_{jk}(h,w,t)| \geq 1 \ \ (t \geq 0).$$

Due to the well-known result (Marcus and Minc, 1964, Section III.9.2) (see also the Appendix A, Section A1), $F(h,w,t) \geq I$. So we take $m_F(r) = 1$. In addition, let $G(h,w,t) = diag\ [g_j(h,w,t)]_{j=1}^n$, such that

$$0 \leq 2f_{jj} - 7/4 < g_j \leq 2f_{jj} - 5/4 \ \ (g_j = g_j(h,w,t);\ j = 1,2,...). \tag{8.10}$$

Then

$$2(f_{jj} - \sum_{k=1,k\neq j}^{n} |f_{jk}|) - g_j \geq$$

$$2f_{jj} - g_j - 1/4 \geq 5/4 - 1/4 = 1.$$

Again use the well-known result from (Marcus and Minc, 1964, Section III.9.2) (see also the Appendix A, Section A1). It gives us the inequality $\Psi(h,w,t) \geq I$. In addition,

$$2(f_{jj} + \sum_{k=1,k\neq j}^{n} |f_{jk}|) - g_j$$

$$\leq 2f_{jj} - g_j + 1/4 < 7/4 + 1/4 = 2.$$

Hence we have $\Psi(h,w,t) < 2I$. Therefore, under (8.9) and (8.10) the zero solution to equation (8.1) is exponentially stable due to Corollary 8.8.2.

8.9 Proof of Theorem 8.8.1

Let us consider in $\mathbf{R}^n$ the linear problem

$$\ddot{x} + 2A(t)\dot{x} + B(t)x = 0 \ \ (t > 0), \tag{9.1}$$

with real piecewise continuous matrices $A(t)$ and $B(t)$. Assume that there is a constant m_A, such that

$$m_A I \leq A_R(t) \text{ and } 0 \leq B_R(t) \leq m_A(2A_R(t) - m_A I) \ \ (t \geq 0). \tag{9.2}$$

Put

$$T(t) := 2m_A A(t) - B(t).$$

Lemma 8.9.1 *Under condition (9.2) any solution $x(t)$ of problem (9.1), (8.2) satisfies the estimate*

$$(p_0 - m_A^2)\|x(t)\|^2 + \|\dot{x}(t)\|^2 \le e^{2\Lambda t}((p_0 - m_A^2)\|x_0\|^2 + \|x_1\|^2)$$

where

$$p_0 := \sup_{t\ge 0} \|T_R(t)\| + \|T_I(t)\| \text{ and } \Lambda := -m_A + \sqrt{p_0 - m_A^2}.$$

Proof: Put in (9.1)

$$x(t) = e^{-mt}y(t) \ \ (m \equiv m_A). \tag{9.3}$$

Then we have the equation

$$\ddot{y} - 2(m - A(t))\dot{y} - C(t)y = 0$$

with

$$C(t) = 2A(t)m - Im^2 - B(t) = T(t) - m^2 I.$$

Reduce this equation to the system

$$\dot{y}_1 = 2(m - A(t))y_1 + C(t)y_2, \ \dot{y}_2 = y_8.$$

Doing the scalar product, we get

$$\frac{d}{dt}(y_1, y_1) = 4((m - A(t))y_1, y_1) + 2(C(t)y_2, y_1),$$

$$\frac{d}{dt}(y_2, y_2) = 2(y_2, y_1).$$

Since the matrices are real, we can write out

$$((mI - A(t))y_1, y_1) = ((mI - A_R(t))y_1, y_1) \le 0.$$

So

$$d(y_1, y_1)/dt \le 2\|C(t)\|\|y_2\|\|y_1\|.$$

Thus

$$d\|y_1(t)\|/dt \le \|C(t)\|\|y_2(t)\|, \ d\|y_2(t)\|/dt = \|y_1(t)\|. \tag{9.4}$$

Furthermore,

$$\|C\| \le \|C_R\| + \|C_I\| = \|T_R - m^2 I\| + \|T_I\| \ \ (C = C(t), T = T(t)).$$

Under the hypothesis of the present lemma C_R is positive. So

$$\|T_R - m^2 I\| = \sup_{v\in R^n, \|v\|=1} (T_R v, v) - m^2 = \|T_R\| - m^2.$$

Thus

$$\|C\| \le \|C_R\| + \|C_I\| = \|T_R - m^2 I\| + \|T_I\| =$$
$$\|T_R\| + \|T_I\| - m_A^2 \le p_0 - m_A^2$$
$$(t \ge 0;\ C = C(t), T = T(t)).$$

Put

$$b_0 := \sqrt{p_0 - m^2}.$$

Then (9.4) implies

$$d\|y_1(t)\|/dt \le b_0^2 \|y_2(t)\|,\ d\|y_2(t)\|/dt = \|y_1(t)\|.$$

Hence,

$$\|y_k(t)\| \le z_k(t)\ \ (k = 1, 2;\ t \ge 0), \tag{9.5}$$

where $(z_1(t), z_2(t))$ is a solution of the coupled system

$$\dot{z}_1 = b_0^2 z_2,\ \dot{z}_2 = z_1\ \ (z_1(0) = \|x_1\|, z_2(0) = \|x_0\|).$$

Simple calculations show that

$$z_1(t) = \frac{1}{2}[(z_1(0) + z_2(0)b_0)e^{b_0 t} + (z_1(0) - z_2(0)b_0)e^{-b_0 t}],$$

$$z_2(t) = \frac{1}{2}[(z_1(0)/b_0 + z_2(0))e^{b_0 t} - (z_1(0)/b_0 - z_2(0))e^{-b_0 t}].$$

By the Schwarz inequality,

$$z_1^2(t) \le\ e^{2b_0 t}\ [(z_1(0) + z_2(0)b_0)^2 + (z_1(0) - z_2(0)b_0)^2] \le$$
$$e^{2b_0 t}\ (z_1^2(0) + z_2^2(0)b_0^2)$$

and

$$z_2^2(t)b_0^2 \le e^{2b_0 t}\ (z_1^2(0) + z_2^2(0)b_0^2).$$

Hence (9.3) and (9.5) yield the required result. □

Proof of Theorem 8.8.1: Consider equation (8.1) as (9.1) with

$$A(t) = F(x(t), \dot{x}(t), t), B(t) = G(x(t), \dot{x}(t), t).$$

If the initial vectors satisfy condition (8.6), then for some $t_0 > 0$,

$$(x(t), \dot{x}(t)) \in \omega_r\ (t \le t_0).$$

Due to Lemma 8.9.1, we have estimate (8.7) for $t \le t_0$. It implies

$$\|x(t)\|^2 + \|\dot{x}(t)\|^2 \le$$
$$e^{2\Lambda_0(r)t} max\ \{\zeta_0(r) - m_F^2(r),\ \frac{1}{\zeta_0(r) - m_F^2(r)}\}(\|x_0\|^2 + \|x_1\|^2)$$
$$(t \le t_0).$$

So according to (8.6) the solution remains in ω_r and we can thus extend (8.7) to all $t \ge 0$, as claimed. □

8.10 Scalar Equations with Real Characteristic Roots

Let us consider the scalar equation

$$y^{(n)} + p_1(y, ..., y^{(n-1)}, t)y^{(n-1)} + ... + p_n(y, ..., y^{(n-1)}, t)y = 0 \ (t \geq 0) \quad (10.1)$$

whose coefficients

$$p_j(h_1, ..., h_n, t) = p_j(h, t) \ \ (h = (h_k) \in \mathbf{R}^n, \ j = 1, ..., n)$$

are real scalar-valued functions continuous in $h = (h_k) \in \mathbf{R}^n$ and piecewise continuous in $t \geq 0$. Introduce the algebraic equation

$$\lambda^n + p_1(h, t)\lambda^{n-1} + ... + p_n(h, t) = 0 \ \ (h \in \mathbf{R}^n; \ t \geq 0). \quad (10.2)$$

Theorem 8.10.1 *Suppose that, for all $h \in \mathbf{R}^n$ and a sufficiently large $t_0 > 0$, the roots*

$$\lambda_1(h, t), ..., \lambda_n(h, t)$$

of the polynomial (10.2) are real and satisfy the inequalities

$$\nu_0 \leq \lambda_1(h, t) \leq \nu_1 \leq \lambda_2(h, t) \leq ... \leq \nu_{n-1} \leq \lambda_n(h, t) \leq \nu_n < 0$$

$$(h \in \mathbf{R}^n; \ t \geq t_0),$$

where ν_i are some constants, such that

$$\nu_0 < \nu_1 < ... < \nu_n < 0.$$

Then equation the zero solution to equation (10.1) is exponentially stable.

For the proof see (Levin, 1969).

9. Lur'e Type Systems

9.1 Stability Conditions

In the present chapter we investigate the asymptotic stability of the system

$$\ddot{w} + B\dot{w}(t) + Cw(t) = F(w, t) \quad (t > 0), \tag{1.1}$$

where B and C are real constant $n \times n$-matrices ($n > 1$); $F(h, t)$ continuously maps $\mathbf{C}^n \times [0, \infty)$ into $\mathbf{C}^n$. Take the initial conditions

$$w(0) = u_0, \ \dot{w}(0) = u_1 \quad (u_0, u_1 \in \mathbf{C}^n). \tag{1.2}$$

A solution of problem (1.1), (1.2) is a function $w : [0, \infty) \to \mathbf{C}^n$, having the first and second locally bounded derivatives and satisfying (1.2) and (1.1) for all $t \geq 0$. Existence of solutions is assumed. Note that the approach suggested in the present chapter allows us to consider the systems

$$\sum_{k=0}^{m} A_k w^{(k)}(t) = F(w, t) \quad (m \geq 2, t \geq 0),$$

where A_k are constant matrices.

Assume that there is a constant $q = q(r) \geq 0$, such that

$$\|F(h, t)\| \leq q\|h\| \quad (h \in \Omega(r), t \geq 0). \tag{1.3}$$

Recall that $\|.\|$ is the Euclidean norm and $\Omega(r) = \{z \in \mathbf{C} : \|z\| \leq r\}$ for a positive $r \leq \infty$; $N(.)$ is the Frobenius norm (see Section 1.1).

Consider the matrix pencil

$$W(z) = Iz^2 + Bz + C \quad (z \in \mathbf{C}). \tag{1.4}$$

It is assumed that pencil $W(z)$ *is stable.* That is, the determinant *det* $(W(z))$ of $W(z)$ is a Hurwitz polynomial.

Clearly, the integrals

$$\psi_k := [\frac{1}{2\pi} \int_{-\infty}^{\infty} y^{2k} \|W^{-1}(iy)\| dy]^{1/2} \quad (k = 0, 1)$$

converge.

Theorem 9.1.1 *Let pencil $W(.)$ be stable, and the conditions (1.3) and*

$$b(W) := \sup_{y \in \mathbf{R}} \|W^{-1}(iy)\| < \frac{1}{q} \tag{1.5}$$

hold. Then the zero solution to equation (1.1) is asymptotically stable. Moreover, with the notations $\eta_B(\tilde{w}_0) = \psi_1\|u_0\| + \psi_0\|u_1 + Bu_0\|$ and

$$\eta_C(\tilde{w}_0) = \psi_1\|u_1\| + \psi_0\|Cu_0\| \quad (\tilde{w}_0 = (u_0, u_1)),$$

all vectors $\tilde{w}_0 = (u_0, u_1) \in \mathbf{C}^{2n}$, satisfying the inequality

$$\zeta(q, \tilde{w}_0) := (2\eta_B(\tilde{w}_0)\eta_C(\tilde{w}_0))^{1/2} + \frac{q\psi_0\eta_B(\tilde{w}_0)}{1 - qb(W)} < r \tag{1.6}$$

belong to the region of attraction of the zero solution to equation (1.1). Besides, any solution w of problem (1.1), (1.2) satisfies the inequalities

$$\|w(t)\| \leq \zeta(q, \tilde{w}_0) \ \ (t \geq 0) \tag{1.7}$$

and

$$\left[\int_0^\infty \|w(t)\|^2 dt\right]^{1/2} \leq \frac{\eta_B(\tilde{w}_0)}{1 - qb(W)}. \tag{1.8}$$

The proof of this theorem is divided into a series of lemmas, which are presented in the next two sections.

Note that the arguments of the proof of Lemma 5.3.3 allow us *to replace $b(W)$ by*

$$\sup_{|y| \leq R_0} \|W^{-1}(iy)\| \text{ where } R_0 := \|B\|/2 + \sqrt{\|B\|^2/4 + 2\|C\|}.$$

Put

$$\phi(W(z)) := \frac{N^{n-1}(W(z))}{(n-1)^{(n-1)/2}|det\ (W(z))|}.$$

Due to Lemma 1.5.7 $\|W^{-1}(z)\| \leq \phi(W(z))$. Thus $\psi_k \leq \tilde{\psi}_k$, where

$$\tilde{\psi}_k := \left[\frac{1}{2\pi}\int_{-\infty}^{\infty} y^{2k}\phi^2(W(iy))dy\right]^{1/2} \ \ (k = 0, 1).$$

Below we will give simple estimates for $\tilde{\psi}_k$ $(k = 0, 1)$. Thus the previous theorem implies

Corollary 9.1.2 *Let pencil $W(.)$ be stable, and the conditions (1.3) and*

$$\tilde{b}(W) := \sup_{y \in \mathbf{R}} \phi(W(iy)) < \frac{1}{q}$$

hold. Then the zero solution to equation (1.1) is asymptotically stable. Moreover, with the notations

$$\tilde\eta_B(\tilde w_0) = \tilde\psi_1 \|u_0\| + \tilde\psi_0 \|u_1 + Bu_0\|$$

and

$$\tilde\eta_C(\tilde w_0) = \tilde\psi_1 \|u_1\| + \tilde\psi_0 \|Cu_0\|$$

all vectors $\tilde w_0 = (u_0, u_1) \in \mathbf{C}^{2n}$*, satisfying the inequality*

$$\tilde\zeta(q, \tilde w_0) := (2\tilde\eta_B(\tilde w_0)\tilde\eta_C(\tilde w_0))^{1/2} + \frac{q\tilde\psi_0 \eta_B(\tilde w_0)}{1 - q\tilde b(W)} < r$$

belong to the region of attraction of the zero solution to equation (1.1). Besides, any solution of problem (1.1), (1.2) satisfies (1.7) and (1.8) with $\zeta(q, \tilde w_0) = \tilde\zeta(q, \tilde w_0)$*,* $\eta_B = \tilde\eta_B$ *and* $b(W) = \tilde b(W)$*.*

It is simple to check that $det\ (W(z)) = det\ (T - zI)$, where

$$T = \begin{pmatrix} -B & -C \\ I & 0 \end{pmatrix}$$

is the matrix of the linear part of (1.1). Thus if all the eigenvalues of T are real, then

$$|det\ (W(iy))| = |det\ (T - iyI)| \geq |\alpha - iy|^{2n} = (\alpha^2 + y^2)^n,$$

where $\alpha = \alpha(T) = \max\ Re\ \lambda_k(T)$, That is, α is the real part of the extreme right root of $det\ (W(z))$. Thus,

$$\tilde\psi_k^2 \leq \frac{\theta_n^2}{2\pi} \int_{-\infty}^{\infty} y^{2k} \frac{N^{2n-2}(W(iy))}{(\alpha^2 + y^2)^{2n}} dy$$

where $\theta_n = (n-1)^{-(n-1)/2}$. But $N(W(iy)) \leq \sqrt{n} y^2 + |y| N(B) + N(C)$. Therefore,

$$\tilde\psi_k^2 \leq \frac{\theta_n^2}{\pi} \int_0^{\infty} \frac{y^{2k}(\sqrt{n}y^2 + yN(B) + N(C))^{2n-2} dy}{(\alpha^2 + y^2)^{2n}} \quad (k = 0, 1). \tag{1.9}$$

These integrals are simple calculated.

Example 9.1.3 *Consider system (1.1) with the diagonal matrices*

$B = diag\ [b_1, ..., b_n]$, $C = diag\ [c_1, ..., c_n]$ having entries $b_k > 0, c_k > 0$. Then

$$N^2(B) = \sum_{k=1}^n b_k^2;\ N^2(C) = \sum_{k=1}^n c_k^2$$

and

$$det\ W(z) = \prod_{k=1}^n (z^2 + zb_k + c_k).$$

Hence,

$$\alpha(T) = \max_k Re\ \{-b_k/2 + \sqrt{b_k^2/4 - c_k}\ \}.$$

Now we can directly apply Corollary 9.1.2 taken into account. Note that in the general case, one can take bounds for the roots of $det\ (W(z))$ from the papers (Gil', 2003b) and (Gil', 2003c)

9.2 Preliminaries

Recall that $L^2 \equiv L^2([0,\infty),\mathbf{C}^n)$ is the Hilbert space of functions defined on $[0,\infty)$ with values in $\mathbf{C}^n$ and the norm

$$\|f\|_{L^2} = [\int_0^\infty \|f(t)\|^2 dt]^{1/2}$$

and $C([0,\infty),\mathbf{C}^n)$ is the Banach space of continuous functions defined on $[0,\infty)$ with values in $\mathbf{C}^n$ and the sup-norm norm $\|.\|_C$. Put

$$G(t) := \frac{1}{2\pi}\int_{-\infty}^{\infty} e^{iyt} W^{-1}(iy)dy \ \ (t \ge 0)$$

and

$$(G * f)(t) = \int_0^t G(t-s)f(s)ds.$$

Lemma 9.2.1 *Let $W(.)$ be stable. Then*

$$\|G * f\|_{L^2} \le b(W)\|f\|_{L^2} \ \ (f \in L^2).$$

Proof: Due to the Laplace transform and the Parseval equality with the previous lemma taken into account,

$$\|G * f\|^2_{L^2} = \frac{1}{2\pi}\int_{-\infty}^{\infty} \|W^{-1}(iy)\tilde f(iy)\|^2 dy \le$$

$$b^2(W)\frac{1}{2\pi}\int_{-\infty}^{\infty} \|\tilde f(iy)\|^2 dy = b^2(W)\|f\|^2_{L^2},$$

where $\tilde f$ is the Laplace transform to f. As claimed. □

Consider the equation

$$\ddot u + B\dot u(t) + Cu(t) = 0. \tag{2.1}$$

Lemma 9.2.2 *Let $W(.)$ be stable. Then a solution u of problem (2.1), (1.2) satisfies the inequalities $\|u\|_{L^2} \le \eta_B(\tilde w_0)$ and $\|\dot u\|_{L^2} \le \eta_C(\tilde w_0)$. Moreover,*

$$\|u\|^2_C \le 2\eta_B(\tilde w_0)\eta_C(\tilde w_0).$$

Proof: Applying the Laplace transformation to equation (2.1), we have

$$(z^2 + Bz + C)\tilde u(z) = u_1 + zu_0 + Bu_0,$$

where $\tilde u(z)$ is the Laplace transform to u and z is the dual variable. Hence,

$$u(t) = \frac{1}{2\pi}\int_{-\infty}^{\infty} e^{iyt} W^{-1}(iy)(u_1 + Bu_0 + iyu_0)dy =$$

$$\dot{G}(t)u_0 + G(t)(u_1 + Bu_0). \tag{2.2}$$

Therefore, according to (2.1),

$$\dot{u}(t) = \ddot{G}(t)u_0 + \dot{G}(t)(u_1 + Bu_0) =$$

$$(-B\dot{G}(t) - CG(t))u_0 + \dot{G}(t)(u_1 + Bu_0).$$

Take into account that

$$B\dot{G}(t) + CG(t) := \frac{1}{2\pi}\int_{-\infty}^{\infty} e^{iyt}(Biy + C)W^{-1}(iy)dy =$$

$$\frac{1}{2\pi}\int_{-\infty}^{\infty} e^{iyt}W^{-1}(iy)(Biy + C)dy = \dot{G}(t)B + G(t)C. \tag{2.3}$$

Hence,

$$\dot{u}(t) = \dot{G}(t)u_1 - G(t)Cu_0. \tag{2.4}$$

Now (2.2) yields the inequality $\|u\|_{L^2} \leq \eta_B(\tilde{w}_0)$. Morever, (2.4) implies the inequality $\|\dot{u}\|_{L^2} \leq \eta_C(\tilde{w}_0)$. Furthermore, as it is shown in Section 1.13, for any vector-valued function

$$h \in L^2([0,\infty), \mathbf{C}^n) \text{ with the property } \dot{h} \in L^2([0,\infty), \mathbf{C}^n),$$

we have

$$\|h(t)\|^2 \leq 2[\int_t^{\infty} \|h(s)\|^2 ds \int_t^{\infty} \|\dot{h}(s)\|^2 ds]^{1/2}.$$

This inequality implies the required inequality for $\|u\|_C$. □

9.3 Proof of Theorem 9.1.1

Lemma 9.3.1 *Let condition (1.3) hold with $r = \infty$. Then under (1.5), inequalities (1.7) and (1.8) are true.*

Proof: For any $h \in L^2$, (1.3) implies

$$\|F(h,t)\|_{L^2} \leq q\|h\|_{L^2}. \tag{3.1}$$

Rewrite (1.1) as

$$w(t) = u(t) + \int_0^t G(t-s)F(w(s),s)ds, \tag{3.2}$$

where u is a solution of problem (2.1), (1.2). Due to (3.1) and Lemma 9.2.1,

$$\|w\|_{L^2} \leq \|u\|_{L^2} + qb(W)\|w\|_{L^2}.$$

Thus due to Lemma 9.2.2 and condition (1.5),

$$\|w\|_{L^2} \leq \frac{\|u\|_{L^2}}{1 - qb(W)} \leq \frac{\eta_B(\tilde{w}_0)}{1 - qb(W)}.$$

So (1.8) holds. Now (3.1), (3.2), (2.5) and the Schwarz inequality yields

$$\|w\|_C \leq \|u\|_C + q\|G\|_{L^2}\|w\|_{L^2} \leq$$

$$\|u\|_C + q\psi_0\|w\|_{L^2} \leq \|u\|_C + \frac{q\psi_0\eta_B(\tilde{w}_0)}{1 - qb(W)}.$$

Hence, due to Lemma 9.2.3, we get

$$\|w\|_C \leq (2\eta_B(\tilde{w}_0)\eta_C(\tilde{w}_0))^{1/2} +$$

$$\frac{q\psi_0\eta_B(\tilde{w}_0)}{1 - qb(W)} = \zeta(q, \tilde{w}_0).$$

As claimed. □

Proof of Theorem 9.1.1: Let $r < \infty$. Due to (1.6), for a small enough $t_0 > 0$ a solution w of problem (1.1), (1.2) lies in $\Omega(r)$ for all $t \leq t_0$. Applying the reasonings of the proof of the previous lemma, we get the inequality

$$\sup_{0 \leq t \leq t_0} \|w(t)\| \leq \zeta(q, \tilde{w}_0).$$

But under (1.6) this inequality can be extended from $[0, t_0]$ to $[0, \infty)$. This proves the theorem. □

10. Aizerman's Problem for Nonautonomous Systems

10.1 Comparison of the Green Functions

Let $b_k(t)$ $(t \geq 0;\ k = 1, ..., n)$ be real non-negative scalar-valued functions having continuous derivatives up to $n-k$-th order. Consider the gradient-type equation

$$P(D)x(t) = \sum_{k=0}^{n-1} \frac{d^k}{dt^k}(b_{n-k}(t)x(t)) \ (t \geq 0), \tag{1.1}$$

where

$$D \equiv \frac{d}{dt};\ P(\lambda) = \lambda^n + c_1\lambda^{n-1} + ... + c_n \ (c_k \equiv const > 0) \tag{1.2}$$

is a Hurwitzian polynomial.

A solution of (1.1) is a function $x(.)$ defined on $[0, \infty)$, having derivatives up to the n-th order, that are bounded for all finite $t \geq 0$. In addition, $x(.)$ satisfies (1.1) and the corresponding initial conditions.

A scalar valued function $W(t, \tau)$ defined for $t \geq \tau \geq 0$ is *the Green function to equation (1.1)* if it satisfies that equation for $t > \tau$ and the initial conditions

$$\lim_{t\downarrow\tau} \frac{\partial^k W(t,\tau)}{\partial t^k} = 0 \ (k = 0, ..., n-2);\ \lim_{t\downarrow\tau} \frac{\partial^{n-1} W(t,\tau)}{\partial t^{n-1}} = 1. \tag{1.3}$$

Put

$$K(t) = \frac{1}{2\pi}\int_{-\infty}^{\infty} e^{i\omega t} P^{-1}(i\omega)d\omega$$

and

$$G(t,s) = \frac{1}{2\pi}\int_{-\infty}^{\infty} e^{i\omega t} P^{-1}(i\omega)Q(\omega i, s)d\omega \ \ (s, t \geq 0),$$

where

$$Q(z,s) = \sum_{k=1}^{n} b_k(s)z^{n-k} \ \ (z \in \mathbf{C}).$$

Theorem 10.1.1 *Let the conditions*

$$K(t) \geq 0 \tag{1.4}$$

and

$$G(t,s) \geq 0 \ (t \geq s \geq 0) \tag{1.5}$$

hold. Then the Green function $W(.,.)$ *to equation (1.1) is non-negative. Moreover,*

$$W(t,\tau) \geq K(t-\tau) \ \ (t \geq \tau \geq 0). \tag{1.6}$$

The proof of this theorem is presented in the next section.

10.2 Proof of Theorem 10.1.1

Clearly, $K(t)$ is the Green function of the equation

$$P(D)u(t) = 0 \ (t \geq 0). \tag{2.1}$$

That is, K is a solution of (2.1) with the initial conditions

$$K^{(k)}(0) = 0 \ (k = 0, ..., n-2); \ \ K^{(n-1)}(0) = 1. \tag{2.2}$$

Set $W(t,0) = x(t)$. Then $x(t)$ satisfies (1.1) with the same initial conditions

$$x^{(k)}(0) = 0 \ (k = 0, ..., n-2); \ \ x^{(n-1)}(0) = 1. \tag{2.3}$$

Due to the variation of constants formula, one can rewrite equation (1.1) in the form

$$x(t) = K(t) + \sum_{k=1}^{n} \int_0^t K(t-s)(b_k(s)x(s))^{(n-k)} ds. \tag{2.4}$$

Take into account initial conditions (2.2) and (2.3). Then, integrating by parts, we arrive at the relation

$$\int_0^t K(t-s)(b_{n-k}(s)x(s))^{(k)} ds = \int_0^t K^{(k)}(t-s) b_{n-k}(s)x(s) ds.$$

Moreover,

$$K^{(k)}(t) = \frac{1}{2\pi} \int_{-\infty}^{\infty} (\omega i)^k e^{i\omega t} P^{-1}(i\omega) d\omega \ \ (k = 0, ..., n-1). \tag{2.5}$$

Consequently,

$$G(t,s) = K^{(n-1)}(t)b_1(s) + K^{(n-2)}(t)b_2(s) + ... + K(t)b_n(s). \tag{2.6}$$

Thus, from (2.4) it follows

$$x(t) = K(t) + \int_0^t G(t-s,s)x(s)ds. \tag{2.7}$$

Hence, the conditions (1.4) and (1.5) imply

$$x(t) \geq K(t) \geq 0 \ (t \geq 0).$$

So for $\tau = 0$ the theorem is proved. But due to the variation of constants formula,

$$W(t,\tau) = K(t-\tau) + \sum_{k=1}^{n} \int_{\tau}^{t} K(t-s)(b_k(s)W(s,\tau))^{(n-k)}ds. \tag{2.8}$$

Repeating our above reasonings for arbitrary $\tau > 0$, according (2.4) we arrive at the required result. □

10.3 Aizerman's Type Problem

Consider the equation

$$P(D)x(t) = (b_1(t)x(t))^{(n-1)} + ... + b_n(t)x(t) + Q(x(t),t) \ \ (t \geq 0), \tag{3.1}$$

where Q is a continuous function defined on $\mathbf{R} \times [0,\infty)$ and satisfying the condition

$$|Q(s,t)| \leq q|s| \ \ (s \in \mathbf{R}; \ t \geq 0) \tag{3.2}$$

with a constant $q \geq 0$. Let $\|\overline{x}_0\|$ be the norm of the initial vector.

Definition 10.3.1 *We will say that equation (3.1) is absolutely stable in the class of nonlinearities (3.2), if there exists a constant M which does not depend on a concrete form of $Q(.,.)$ (but which depends on q), such that $|x(t)| \leq M \ \|\overline{x}_0\|$ for all $t \geq 0$ and any solution $x(t)$ of (3.1).*

Problem 10.3.1: *To find a class of systems of the type (3.1) that have the following property: for the absolute stability of (3.1) in the class of nonlinearities (3.2), it is sufficient that the linear equation*

$$P(D)x(t) = (b_1(t)x(t))^{(n-1)} + ... + b_n(t)x(t) + q_1x(t) \tag{3.3}$$

be asymptotically stable for some $q_1 \in [0,q]$.

Lemma 10.3.2 *Let the Green function $W(.,.)$ to equation (1.1) be nonnegative. Then equation (3.1) is absolutely stable in the class of nonlinearities (3.2) provided (3.3) is asymptotically stable with $q_1 = 0$ and $q_1 = q$.*

Proof: Equation (3.1) is equivalent to the equation

$$x(t) = \tilde{x}(t) + \int_0^t W(t,s)Q(x(s),s)ds$$

where $\tilde{x}(t)$ is a solution of (1.1). Clearly, (3.3) takes the form (1.1) when $q_1 = 0$. Thus,

$$|x(t)| \leq |\tilde{x}(t)| + q\int_0^t |W(t,s)||x(s)|ds. \tag{3.4}$$

Since (1.1) is stable and the Green function is positive, we have

$$|\tilde{x}(t)| \leq c_2 \equiv const\ \|\overline{x}_0\|$$

and

$$|x(t)| \leq c_2 + q\int_0^t W(t,s)|x(s)|ds.$$

Due to Lemma 1.7.1, $|x(t)| \leq w(t)$ where w is the solution of the equation

$$w(t) = c_2 + q\int_0^t W(t,s)w(s)ds.$$

But according to the variation of constants formula, this equation is equivalent to (3.3) with $q_1 = q$. Moreover,

$$w(0) = c_2,\ w^{(k)}(0) = 0\ (k = 1, ..., n-1).$$

Since (3.3) with $q_1 = q$ is stable, we have

$$|x(t)| \leq w(t) \leq c_3\, c_2 = M\|\overline{x}_0\|\ \ (c_3 \equiv const;\ t \geq 0).$$

As claimed. □

The latter lemma and Theorem 10.1.1 imply

Theorem 10.3.3 *Let conditions (1.4) and (1.5) hold. Then equation (3.1) is absolutely stable in the class of nonlinearities (3.2), provided (3.3) is asymptotically stable with $q_1 = 0$ and $q_1 = q$.*

Clearly, the latter theorem separates a class of systems satisfying Problem 10.3.1.

Below in Section 10.6 we suggest the explicit absolute stability conditions.

10.4 Equations with Purely Real Roots

Let polynomial $P(\lambda)$ defined by (1.2) have the real roots

$$\beta = \lambda_1 \le \lambda_2 \le .... \le \lambda_n = \alpha < 0. \tag{4.1}$$

Then Lemma 1.11.2 gives us the inequality

$$K(t) \ge \frac{t^{n-1}e^{\beta t}}{(n-1)!}\ (t \ge 0). \tag{4.2}$$

Moreover, thanks to Lemma 1.11.2, the condition

$$\frac{\partial^{n-1} e^{zt} Q(z,s)}{\partial z^{n-1}} \ge 0\ (\lambda_1 \le z \le \lambda_n;\ \ t, s \ge 0)$$

implies (1.5). But $e^{\theta t} \ge 0$ for any real θ. So, if

$$\frac{\partial^j Q(z,s)}{\partial z^j} = \sum_{k=1}^{n-j} \frac{(n-k)! b_k(s) z^{n-k-j}}{(n-k-j)!} \ge 0$$

$$(j = 0, ..., n-1;\ \lambda_1 \le z \le \lambda_n;\ \ s \ge 0), \tag{4.3}$$

then relation (1.5) is valid. Now Theorem 10.1.1 implies

Theorem 10.4.1 *Let polynomial $P(\lambda)$ have the purely real roots (4.1). In addition, let condition (4.3) hold. Then the Green function $W(.,.)$ to equation (1.1) is non-negative. Moreover,*

$$W(t,\tau) \ge K(t-\tau) \ge \frac{(t-\tau)^{n-1} e^{\beta(t-\tau)}}{(n-1)!}\ \ (t \ge \tau \ge 0).$$

Let $n = 2$ and the polynomial

$$P(z) = z^2 + c_1 z + c_2$$

have real roots $\lambda_1 \le \lambda_2$. In this case

$$Q(s,z) \equiv b_1(s) z + b_2(s)\ \ (\lambda_1 \le z \le \lambda_2).$$

In addition,

$$G(t,s) = \dot{K}(t) b_1(s) + K(t) b_2(s)$$

and

$$K(t) = (\lambda_2 - \lambda_1)^{-1}(e^{\lambda_2 t} - e^{\lambda_1 t}) \text{ if } \lambda_1 < \lambda_2$$
$$\text{and } K(t) = t e^{\lambda_1 t} \text{ if } \lambda_1 = \lambda_2. \tag{4.4}$$

So according to (4.3), the conditions

$$Q'_z(s,z) \equiv b_2(s) \ge 0 \text{ and } b_1(s) z + b_2(s) \ge 0\ \ (\lambda_1 \le z \le \lambda_2)$$

imply (1.5). Rewrite the latter relations as

$$b_2(s) \ge 0 \text{ and } b_1(s)\lambda_1 + b_2(s) \ge 0\ \ (s \ge 0). \tag{4.5}$$

Now Theorem 10.4.1 yields

Corollary 10.4.2 *Let the polynomial $P(z) = z^2 + c_1 z + c_2$ have real roots $\lambda_1 \le \lambda_2$ and condition (4.5) hold. Then the Green function $W(.,.)$ of the equation*

$$\ddot{x} + c_1\dot{x} + c_2 x = d(b_1(t)x)/dt + b_2(t)x \quad (x = x(t)) \tag{4.6}$$

is non-negative. Moreover,

$$W(t,\tau) \ge K(t-\tau) \ge e^{\lambda_1(t-\tau)}(t-\tau) \quad (t \ge \tau \ge 0),$$

where K is defined by (4.4).

Note that the equation

$$\ddot{x} + a_1(t)\dot{x} + a_2(t)x = 0$$

with bounded functions $a_1(t), a_2(t)$ can be written as (4.6), if we take

$$c_k = max_{t\ge 0} a_k(t) \ (k = 1,2), \ b_1(t) = a_1(t) - c_1,$$

$$b_2(t) = a_2(t) - c_2 - a_1'(t),$$

provided $a_1(t)$ is differentiable. Consider now the third order equation

$$\frac{d^3x}{dt^3} + c_1\frac{d^2x}{dt^2} + c_2\frac{dx}{dt} + c_3 x = (b_2(t)x)' + b_3(t)x \quad (t \ge 0). \tag{4.7}$$

So $b_1(t) \equiv 0$ and $Q(z,s) = b_2(s)z + b_3(s)$. Let the polynomial

$$P(z) = z^3 + c_1 z^2 + c_2 z + c_3 \tag{4.8}$$

have the real roots

$$\lambda_1 \le \lambda_2 \le \lambda_3. \tag{4.9}$$

Then the inequalities

$$b_2(s) \ge 0 \text{ and } b_3(s) + b_2(s)\lambda_1 \ge 0 \ (s \ge 0) \tag{4.10}$$

provide condition (4.3). Now Theorem 10.4.1 yields

Corollary 10.4.3 *Let conditions (4.9) and (4.10) hold. Then the Green function $W(.,.)$ to equation (4.7) is non-negative. Moreover,*

$$W(t,\tau) \ge \frac{(t-\tau)^2}{2} e^{\lambda_1(t-\tau)} \ (t \ge \tau \ge 0).$$

10.5 Equations with Nonreal Roots

Recall the following result (see Section 6.1): let the polynomial defined by (4.8) have a pair of complex conjugate roots: $-\gamma \pm i\omega$, and a real root $-z_0$ $(z_0, \gamma, \omega > 0)$. Then the function

$$K(t) \equiv \frac{1}{2\pi i} \int_{-i\infty}^{i\infty} \frac{exp(t\lambda)d\lambda}{\lambda^3 + c_1\lambda^2 + c_2\lambda + c_3} \tag{5.1}$$

is non-negative provided that

$$\gamma > z_0. \tag{5.2}$$

Now Theorem 10.1.1 implies

Corollary 10.5.1 *Let the polynomial defined by (4.8) have a pair of complex conjugate roots: $-\gamma \pm i\omega$, and a real root $-z_0$ $(z_0, \gamma, \omega > 0)$. In addition, let condition (5.2) hold. Then the Green function $W(t, \tau)$ to the equation*

$$\frac{d^3x(t)}{dt^2} + c_1\frac{d^2x(t)}{dt^2} + c_2\frac{dx(t)}{dt} + c_3x(t) - b_3(t)x(t) = 0 \; (b_3(t) \geq 0, \; t \geq 0)$$

is non-negative. Moreover, inequality (1.6) holds, where K is defined by (5.1).

Indeed, in this case $n = 3$ and $b_1(t) \equiv b_2(t) \equiv 0$. We need only the condition $K(t) \geq 0$. Now Theorem 10.1.1 imply the required result.

Since the Laplace transform of a convolution is a product of the Laplace transforms, we can use the results of this section in the case $n > 3$.

10.6 Absolute Stability Conditions

10.6.1 Estimates for Green's Functions

In the present section the positivity of K and $G(.,.)$ is not assumed. To establish stability conditions, wee need some estimates for Green's function of equation (1.1).

Let $\lambda_1, ..., \lambda_n$ be the roots of $P(\lambda)$ taken with their multiplicities,

$$\alpha = max_k Re\ \lambda_k \text{ and } \Lambda = max_k|\lambda_k|.$$

Due to Lemma 5.7.1,

$$|K^{(j)}(t)| \leq e^{\alpha t}\eta_j(t) \;\; (t \geq 0), \tag{6.1}$$

where

$$\eta_j(t) = \sum_{k=0}^{n-1} \frac{j!\Lambda^{j-k}t^{n-k-1}}{(j-k)!(n-k-1)!k!}.$$

In particular,

$$|K(t)| \leq \frac{t^{n-1}e^{\alpha t}}{(n-1)!} \quad (t \geq 0).$$

Moreover, from Corollary 5.7.3 it follows,

$$\int_0^\infty |K^{(j)}(t)|\, dt \leq \frac{(\Lambda + |\alpha|)^j}{|\alpha|^n}. \tag{6.2}$$

Lemma 10.6.1 *Let*

$$m_k := \sup_{t\geq 0} |b_k(t)| < \infty \; (k = 1, ..., n) \tag{6.3}$$

and

$$\gamma_0 := \frac{1}{|\alpha|^n} \sum_{j=0}^{n-1} m_j \, (\Lambda + |\alpha|)^{n-j-1} < 1. \tag{6.4}$$

Then the Green function to equation (1.1) satisfies the estimate

$$\max_{0\leq\tau\leq t\leq\infty} |W(t,\tau)| \leq \frac{l_n(\alpha)}{1-\gamma_0},$$

where

$$l_n(\alpha) = [(n-1)!]^{-1} \max_{t\geq 0} e^{\alpha t} t^{n-1} = \frac{(n-1)^{n-1}}{(n-1)! e^{n-1} |\alpha|^{n-1}}.$$

Proof: By (2.6) and (6.3), we have

$$|G(t,s)| \leq |K^{(n-1)}(t)| m_1 + |K^{(n-2)}(t)| m_2 + .. + |K(t)| m_n \; (t, s \geq 0).$$

With (6.1) taken into account, we get

$$|G(t,s)| \leq e^{\alpha t}(\eta_{n-1}(t) m_1 + ... + \eta_0(t) m_n). \tag{6.5}$$

According to (2.8),

$$|W(t,\tau)| \leq |K(t-\tau)| + \int_\tau^t [|K^{(n-1)}(t-s)| m_1 + ...$$

$$+ |K(t-s)| m_n] |W(s,\tau)| ds. \tag{6.6}$$

So,

$$\max_{t\geq\tau\geq 0} |W(t,\tau)| \leq \max_{t\geq\tau\geq 0} |K(t-\tau)| + \gamma \max_{t\geq\tau\geq 0} |W(t,\tau)|, \tag{6.7}$$

where

$$\gamma \equiv \int_0^\infty [\, |K^{(n-1)}(s)| m_1 + ... + |K(s)| m_n \,] ds.$$

But due to (2.6) $\gamma \leq \gamma_0$. In addition, (6.1) implies

$$\max_{t\geq 0}|K(t)| \leq l_n(\alpha).$$

Thus,

$$\max_{t\geq\tau\geq 0}|W(t,\tau)| \leq l_n(\alpha) + \gamma_0 \max_{t\geq\tau\geq 0}|W(t,\tau)|.$$

Then, thanks to (6.4), we get the required result. □

Lemma 10.6.2 *Under conditions (6.3) and (6.4) the Green function to equation (1.1) satisfies the inequality*

$$\max_{t\geq 0}\int_0^t |W(t,\tau)|d\tau \leq \frac{1}{(1-\gamma_0)\,|\alpha|^n}.$$

Proof: According to (6.6),

$$\int_0^t |W(t,\tau)|d\tau \leq \int_0^t |K(t-\tau)|d\tau + \int_0^t\int_\tau^t [|K^{(n-1)}(t-s)|m_1+$$

$$... + |K(t-s)|m_n]|W(s,\tau)|ds\, d\tau. \tag{6.8}$$

But

$$\int_0^t\int_\tau^t |K^{(k)}(t-s)|\,|W(s,\tau)|\,ds\,d\tau = \int_0^t\int_0^s |K^{(k)}(t-s)||W(s,\tau)|d\tau\, ds$$

$$\leq \int_0^t |K^{(k)}(t-s)|ds\; max_{s\geq 0}\int_0^s |W(s,\tau)|d\tau \leq$$

$$\int_0^\infty |K^{(k)}(s)|ds\; max_{s\geq 0}\int_0^s |W(s,\tau)|d\tau.$$

Taking into account (6.2), from (6.8) we get

$$\max_{t\geq 0}\int_0^t |W(t,\tau)|d\tau \leq |\alpha|^{-n} + \gamma_0 \max_{t\geq 0}\int_0^t |W(t,\tau)|d\tau.$$

Now (6.4) yields the required result. □

10.6.2 Absolute Stability Conditions

Theorem 10.6.3 *Let the conditions (6.3), (6.4) and*

$$\frac{q}{(1-\gamma_0)\,|\alpha|^n} < 1 \tag{6.9}$$

hold. Then equation (3.1) is absolutely stable in the class of nonlinearities (3.2).

Proof: Inequality (3.4) and Lemma 10.6.2 imply

$$\sup_{t\geq 0}|x(t)| \leq \sup_{t\geq 0}|\tilde{x}(t)| + q\sup_{s\geq 0}|x(s)|\frac{1}{(1-\gamma_0)\,|\alpha|^n}.$$

Now the result is due to condition (6.9). □

10.7 Positive Solutions of Nonlinear Equations

Consider the equation

$$P(D)x(t) = (b_1(t)x(t))^{(n-1)} + ... + b_n(t)x(t) +$$

$$F(x(t), \dot{x}(t), ..., x^{(n-1)}(t), t) \ \ (t \geq 0), \tag{7.1}$$

where F is a scalar valued continuous function defined on $\mathbf{R}^n \times [0, \infty)$.

Denote

$$R^n_+ \equiv \{\overline{y} \equiv column\ (y_1, ..., y_n) \in \mathbf{R}^n : \ y_1 \geq 0\}.$$

That is, R^n_+ is the subset of vectors whose first coordinate is non-negative.

Lemma 10.7.1 *Let the Green function $W(t,\tau)$ to (1.1) be non-negative and*

$$F(\overline{y}, t) \equiv F(y_1, y_2, ..., y_n, t) \geq 0 \ \ (\overline{y} \in R^n_+, t \geq 0). \tag{7.2}$$

Then equation (7.1) possesses non-negative solutions.

Proof: Due to the variation of constants formula, equation (7.1) with the initial conditions (2.3) can be written in the form

$$x(t) = W(t,0) + \int_0^t W(t,s)F(\overline{x}(s), s)ds,$$

where

$$\overline{x}(t) = column\ (x(t), \dot{x}(t), ..., x^{(n-1)}(t)).$$

Since, $W(t,0) \geq 0$, for a sufficiently small t_0 we have $x(t) \geq 0$ $(t \leq t_0)$. So

$$x(t) \geq W(t,0) \geq 0 \ (t \leq t_0).$$

Extending this relation to all $t \geq 0$, we arrive at the required result. □

The latter lemma and Theorem 10.1.1 yields

Theorem 10.7.2 *Let conditions (1.4), (1.5) and (7.2) hold. Then equation (7.1) possesses non-negative solutions.*

11. Input-to-State Stability

11.1 Definitions and Preliminaries

Let us consider a system described by the following equation in a Euclidean space $\mathbf{C}^n$:

$$\dot{x}(t) = \Phi(x(t), u(t), t) \quad (t \geq 0), \tag{1.1}$$

where $x : R_+ \to \mathbf{C}^n$ is the state, $u : R_+ \to \mathbf{C}^m$ is the input and Φ maps $\mathbf{C}^n \times \mathbf{C}^m \times R_+$ into $\mathbf{C}^n$ with the property $\Phi(0,0,t) \equiv 0$. Here $R_+ = [0, \infty)$.

In the present chapter it is convenient for us to denote by $\|.\|_{C^n}$ *the Euclidean norm in* $\mathbf{C}^n$. *In addition,* $\|w\|_{L^p,n}$ *is the norm in* $L^p(R_+, \mathbf{C}^n)$ $(1 \leq p \leq \infty)$. That is,

$$\|w\|_{L^p,n} = [\int_0^\infty \|w(t)\|^p_{C^n} dt]^{1/p} \quad (1 \leq p < \infty;\ w \in L^p(R_+, \mathbf{C}^n))$$

and

$$\|w\|_{L^\infty,n} = ess \sup_{t \geq 0} \|w(t)\|_{C^n} \quad (w \in L^\infty(R_+, \mathbf{C}^n))$$

(see also Section 1.13). Let X and U be Banach spaces of vector-valued functions defined on R_+ with values in $\mathbf{C}^n$ and $\mathbf{C}^m$, respectively. For instance, $X = L^p(R_+, \mathbf{C}^n)$ $(1 \leq p \leq \infty)$ and $U = L^{p_1}(R_+, \mathbf{C}^m)$ $(1 \leq p_1 \leq \infty)$. Let $\|.\|_X$ and $\|.\|_U$ be the norms in X and U, respectively.

Definition 11.1.1 *System (1.1) is said to be input-to-state (UX)-stable, if for any* $\epsilon > 0$, *there is a* $\delta > 0$, *such that the conditions* $\|u\|_U \leq \delta$ *and*

$$x(0) = 0 \tag{1.2}$$

imply $\|x\|_X \leq \epsilon$. *System (1.1) is said to be globally input-to-state (UX)-stable if the conditions (1.2) and* $u \in U$ *imply* $x \in X$.

Let us consider the system

$$\dot{x}(t) = B(x(t), t)x(t) + F(x(t), u(t), t), \tag{1.3}$$

where $x : [0, \infty) \to \mathbf{C}^n$ is the state, $u : [0, \infty) \to \mathbf{C}^m$ is the input, again. Besides, $B(h, t)$ for each $h \in \mathbf{C}^n$ is a variable $n \times n$-matrix continuously dependent on its arguments, and F continuously maps $\mathbf{C}^n \times \mathbf{C}^m \times [0, \infty)$ into $\mathbf{C}^n$. It is assumed that there are constants $\nu_S, \nu_I \geq 0$ such that

$$\|F(h, z, t)\|_{C^n} \leq \nu_S \|h\|_{C^n} + \nu_I \|z\|_{C^m} \quad (h \in \mathbf{C}^n;\ z \in \mathbf{C}^m,\ t \geq 0) \tag{1.4}$$

Denote by $W_v(t, s)$ the evolution operator of the equation

$$\dot{x}(t) = B(v(t), t)x(t) \tag{1.5}$$

for an arbitrary differentiable function $v : R_+ \to \mathbf{C}^n$.

Theorem 11.1.2 *Let the conditions (1.4) and*

$$\kappa_0 := \sup_{v,t} \int_0^t \|W_v(t, s)\| ds < \frac{1}{\nu_S}$$

hold, where the supremum is taken over all $t \in R_+$ and differentiable functions $v : R_+ \to C^n$. Then with

$$X = L^\infty(R_+, \mathbf{C}^n) \text{ and } U = L^\infty(R_+, \mathbf{C}^m)$$

system (1.3) is globally input-to-state (UX)-stable.

Proof: Rewrite equation (1.3) in the form

$$x(t) = \int_0^t W_x(t, s)F(x(s), u(s), s)]ds.$$

Hence

$$\|x\|_{L^\infty, n} \leq \nu_S \ \|x\|_{L^\infty, n} \sup_{t \geq 0} \int_0^t \|W_x(t, s)\|_{C^n} ds + c(u),$$

where

$$c(u) = \nu_I \sup_{t \geq 0} \int_0^t \|W_x(t, s)\|_{C^n} \|u(s)\|_{C^m} ds.$$

This proves the stated result. The details are left to the reader. □

11.2 Systems with Time-Variant Linear Parts

Let us consider the system

$$\dot{x}(t) = A(t)x(t) + F(x(t), u(t), t), \tag{2.1}$$

where $x : [0, \infty) \to \mathbf{C}^n$ is the state, $u : [0, \infty) \to \mathbf{C}^m$ is the input, again. Besides, $A(t)$ is a variable $n \times n$-matrix and F continuously maps $\mathbf{C}^n \times \mathbf{C}^m \times [0, \infty)$ into $\mathbf{C}^n$. It is assumed that

$$\|A(t) - A(s)\| \leq q\,|t - s|\ (t, s \geq 0;\ q = const > 0). \tag{2.2}$$

In addition,

$$v = \sup_{t \geq 0} g(A(t)) < \infty, \text{ and } \rho \equiv -\sup_{t \geq 0} \alpha(A(t)) > 0. \tag{2.3}$$

and

$$\sum_{j=0}^{n-1} \frac{v^j}{\sqrt{j!}} \left[\frac{q(j+1)}{\rho^{j+2}} + \frac{\nu_S}{\rho^{j+1}}\right] < 1. \tag{2.4}$$

Recall that $g(A)$ and $\alpha(A)$ are defined in Sections 1.5 and 1.3, respectively.

Theorem 11.2.1 *Let conditions (1.4) and (2.2)-(2.4) hold. Then with*

$$X = L^\infty(R_+, \mathbf{C}^n) \text{ and } U = L^\infty(R_+, \mathbf{C}^m)$$

system (2.1) is globally input-to-state (UX)-stable.

The proof of this theorem is presented in the next section.

Example 11.2.2 *Consider the one contour system with a scalar input*

$$\ddot{y} + a(t)\dot{y} + b(t)y = f(t, y, \dot{y}) + u(t), \tag{2.5}$$

where $a(t), b(t)$ are positive scalar-valued bounded functions with the property

$$|a(t) - a(s)| + |b(t) - b(s)| \leq q_0|t - s| \tag{2.6}$$

for all $t, s \geq 0$. Besides, it is assumed that the scalar-valued function $f : [0, \infty) \times \mathbf{R}^2 \to \mathbf{R}$ satisfies the condition

$$|f(t, y, z)|^2 \leq \nu_S^2(|y|^2 + |z|^2) \tag{2.7}$$

for all $t \geq 0$; $y, z \in \mathbf{R}$. Due to Example 1.6.1, we easily get

$$v \leq 1 + \sup_{t \geq 0} b(t).$$

Assume that for some $\rho > 0$,

$$\alpha(A(t)) = Re\{-\frac{a(t)}{2} + \sqrt{\frac{a^2(t)}{4} - b(t)}\} \leq -\rho\ (t \geq 0)$$

and

$$\frac{q_0}{\rho^2} + \frac{\nu_S}{\rho} + \frac{2vq_0}{\rho^3} + \frac{\nu_S}{\rho^2} < 1.$$

Then due to Theorem 11.2.1, system (2.5) under conditions (2.7) and (2.6) is globally input-to-state (UX) - stable with $X = U = L^\infty(R_+)$.

11.3 Proof of Theorem 11.2.1

By Corollary 1.5.3 we have

$$\|exp[A(\tau)t]\|_{C^n} \leq \Gamma(t, A(.)) \ (t, \tau \geq 0), \tag{3.1}$$

where

$$\Gamma(t, A(.)) = e^{-\rho t} \sum_{k=0}^{n-1} \frac{v^k t^k}{(k!)^{3/2}}.$$

Rewrite equation (2.1) in the form

$$\dot{x} - A(\tau)x = [A(t) - A(\tau)]x + F(x, u, t), \tag{3.2}$$

regarding arbitrary $\tau \geq 0$ as fixed. With $x(0) = 0$, this equation is equivalent to the following one:

$$x(t) = \int_0^t exp[(A(\tau)(t-s)][(A(s)-$$

$$A(\tau))x(s) + F(x(s), u(s), s)]ds.$$

Due to (3.1) and (3.2) the latter equality implies the relation

$$\|x(t)\|_{C^n} \leq \int_0^t \Gamma(t-s, A(.))\times$$

$$\{[q|\tau - s| + \nu_S\|x(s)\|_{C^n} + \nu_I\|u(s)\|_{C^m}\}ds.$$

Hence

$$\|x(t)\|_{C^n} \leq \sup_{s \leq t} \|x(s)\|_{C^n} \times$$

$$\int_0^t \Gamma(t-s, A(.))\{q(t-s)ds + \nu_S\}ds + b(u)$$

where

$$b(u) = \nu_I \sup_{t \geq 0} \int_0^t \Gamma(t-s, A(.))\|u(s)\|_{C^m} ds.$$

Simple calculations show that

$$\int_0^t \Gamma(t-s, A(.))(t-s)ds \leq \theta_1(A)$$

with

$$\theta_1(A) = \sum_{k=0}^{n-1} \frac{(k+1)v^k}{\sqrt{k!}\rho^{k+2}}.$$

Moreover,

$$\int_0^t \Gamma(t-s, A(.))ds \leq \theta_0(A) \equiv \sum_{k=0}^{n-1} \frac{\upsilon^k}{\rho^{k+1}\sqrt{k!}} \quad (t \geq 0).$$

Consequently,

$$b(u) \leq \nu_I \theta_0 \; ess \sup_{t\geq 0} \|u(t)\|$$

By (3.2) it can be written

$$\sup_{s\geq 0} \|x(s)\|_{C^n} \leq \sup_{s\geq 0} \|x(s)\|_{C^n} (q\theta_1 + \nu\theta_0) + b(u).$$

Due to (2.4)

$$\sup_{s\leq t} \|x(s)\|_{C^n} \leq \frac{b(u)}{1 - q\theta_1 - \nu\theta_0}.$$

This proves the stated result. □

11.4 The Input-to-State Version of Aizerman's Problem

Consider the system

$$\dot{x}(t) = Ax(t) + b\, f(s(t), u(t), t) \;\; (s(t) = cx(t), \; t \geq 0), \tag{4.1}$$

where $x(t) : R_+ \to \mathbf{R}^n$ is the state, $u(t) : R_+ \to \mathbf{R}^m$ is the input, A is a constant real Hurwitz matrix, b is a real column, c is a real row, f maps $\mathbf{R}^1 \times \mathbf{R}^m \times R_+$ into $\mathbf{R}^1$ with the property: there are constants $q_S, q_I \geq 0$ such that

$$|f(s,h,t)| \leq q_S|s| + q_I\|h\|_{C^m} \text{ for all } s \in \mathbf{R}^1, h \in \mathbf{R}^m \text{ and } t \geq 0. \tag{4.2}$$

In the present section we consider the following version of the Aizerman problem: to separate a class of systems such that the asymptotic stability of the linear system

$$\dot{x}(t) = Ax(t) + b\, q_1 cx \tag{4.3}$$

with some $q_1 \in [0, q_S]$ provides the global input-to-state stability.

Let

$$W(\lambda) = c(\lambda I - A)^{-1} b = \frac{Q(\lambda)}{P(\lambda)}$$

be the transfer function. Here $P(\lambda), Q(\lambda)$ are polynomials. Besides P is monic and Hurwitzian. Recall that

$$K(t) := \frac{1}{2\pi} \int_{-\infty}^{\infty} exp[i\omega t] W(i\omega)\, d\omega$$

is the impulse function.

Theorem 11.4.1 *Let the conditions (4.2) and*

$$K(t) \geq 0 \text{ for all } t \geq 0. \tag{4.4}$$

hold, and let the polynomial $P(\lambda) - q_S Q(\lambda)$ be Hurwitzian. Then with $X = L^\infty(R_+, \mathbf{R})$ and $U = L^\infty(R_+, \mathbf{C}^m)$, system (4.1) is globally input-to-state (UX)-stable.

The proof of this theoren is presented in the next section.

It is simple to check that the Hurwitzness of the polynomials $P(\lambda) - q_S Q(\lambda)$ is equivalent to the asymptotic stability of linear system (4.3) with $q_1 = q_S$.

Due to Lemma 6.1.3, under (4.4), the polynomial $P(z) - q_S Q(z)$ is Hurwitzian, if $P(0) > q_S Q(0) > 0$. Now the previous theorem implies

Corollary 11.4.2 *Let the conditions (4.2), (4.4) and $P(0) > q_S Q(0) > 0$ hold. Then with $X = L^\infty(R_+, \mathbf{R})$ and $U = L^\infty(R_+, \mathbf{C}^m)$, system (4.1) is globally input-to-state (UX)-stable.*

Example 11.4.3 *Consider the following one contour system with a scalar input:*

$$\frac{d^2x}{dt^2} + a_1 \frac{dx}{dt} + a_2 x = b_1 \phi(x, u) + \frac{d\phi(x, u)}{dt}, \tag{4.5}$$

where $\phi(s, u) : R^2 \to R^1$ satisfies the condition

$$|\phi(s, z)| \leq q_S |s| + q_I |z| \ (s, z \in \mathbf{R}). \tag{4.6}$$

Let the polynomial $P(\lambda) = \lambda^2 + a_1 \lambda + a_2$ have real roots $\lambda_1 \leq \lambda_2 < 0$. Then under (4.6) according to the results of Section 6.1, relation (4.4) holds. That is, if

$$P(0) = a_2 > q_S Q(0) = q_S b_1,$$

then by Corollary 11.4.2, system (4.5) is globally input-to-state (UX)-stable, with

$$X = U = L^\infty(R_+, \mathbf{R}),$$

provided (4.6) holds.

Example 11.4.4 *Consider the following one contour system with a scalar input:*

$$\frac{d^3x}{dt^3} + a_1 \frac{d^2x}{dt^2} + a_2 \frac{dx}{dt} + a_3 x = \phi(x, u), \tag{4.7}$$

where $\phi(x, u)$ is the same as in the previous example. Let the polynomial

$$P(\lambda) = \lambda^3 + a_1 \lambda^2 + a_2 \lambda + a_3$$

have negative real roots. Then according to the results of Section 6.1, relation (4.4) holds. Thus, if $a_3 > q_S$, then system (4.7) is globally input-to-state (UX)-stable, with $X = U = L^\infty(R_+, \mathbf{R})$, provided (4.6) holds.

11.5 Proof of Theorem 11.4.1

Equation (4.1) is equivalent to following one:

$$x(t) = exp[At]x(0) + \int_0^t exp[A(t-\tau)]bf(s(\tau), u(\tau), \tau)d\tau.$$

Multiplying this equation by c, and taking into account that $x(0) = 0$, we have

$$s(t) = \int_0^t K(t-\tau)f(s(\tau), u(\tau), \tau)d\tau,$$

since $c\exp[At]b = K(t)$. Due to (4.2), we arrive at the inequality

$$|s(t)| \le \int_0^t K(t-\tau)(q_S|s(\tau)| + q_I\|u(\tau)\|_{C^m})d\tau.$$

Or

$$|s(t)| \le \int_0^t K(t-\tau)q_S|s(\tau)|d\tau + h(t)$$

with the notation

$$h(t) = q_I \int_0^t K(t-\tau)\|u(\tau)\|_{C^m}d\tau.$$

Since A is a Hurwitz matrix, there are constants $\alpha_0, c_1 > 0$ such that

$$0 \le c\exp[At]B = K(t) \le c_1 e^{-\alpha_0 t} \ (t \ge 0).$$

So

$$h(t) \le q_I c_1 \|u\|_{L^\infty,m} \int_0^t e^{-\alpha_0(t-\tau)}d\tau \le b(u) \ (t \ge 0), \tag{5.1}$$

where

$$b(u) = q_I c_1 \alpha_0^{-1} \|u\|_{L^\infty,m}.$$

Consequently,

$$|s(t)| \le \int_0^t K(t-\tau)q_S|s(\tau)|d\tau + b(u).$$

From Lemma 1.7.1 it follows that

$$|s(t)| \le \eta(t), \tag{5.2}$$

where η is the solution of the equation

$$\eta(t) = b(u) + \int_0^t K(t-\tau)q_S\eta(\tau)d\tau.$$

We will solve this equation by the Laplace transformation. Since the transform of the convolution is equal to the product of the transforms, after straightforward calculations and taking into account that $P^{-1}(\lambda)Q(\lambda)$ is the Laplace transform of $K(t)$, we get the equation

$$\overline{\eta}(\lambda) = \frac{b(u)}{\lambda} + P^{-1}(\lambda)Q(\lambda)q_S\overline{\eta}(\lambda),$$

where $\overline{\eta}(\lambda)$ is the Laplace transform of $\eta(t)$. Thus,

$$\eta(t) = \frac{1}{2\pi i}\int_{-i\infty}^{i\infty} \exp[\lambda t](P(\lambda) - q_S Q(\lambda))^{-1}P(\lambda)\lambda^{-1}d\lambda\, b(u).$$

Since $P(\lambda) - q_S Q(\lambda)$ is a Hurwitz polynomial, the function under the integral has no the poles in the right open half-plane, and has one simple pole on the imaginary axis. So, thanks to the residue theorem

$$\eta(t) \leq const\; b(u) \;(t \geq 0).$$

Consequently,

$$\eta(t) \leq const\; \|u\|_{L^\infty,m} \;(t \geq 0).$$

This and (5.2) prove the result. □

12. Orbital Stability and Forced Oscillations

12.1 Global Orbital Stability

Let us consider a system described by the equation

$$\dot{x}(t) = A(t)x(t) + F(x(t), t), \tag{1.1}$$

where $A(t)$ is a variable $n \times n$-matrix, and F maps $\mathbf{C}^n \times [0, \infty)$ into $\mathbf{C}^n$.

Definition 12.1.1 *We will say that system (1.1) is globally exponentially orbitally stable, if there exist constants $N, \epsilon > 0$, such that*

$$\|x_1(t) - x_2(t)\| \leq N\, exp(-\epsilon t)\, \|x_2(0) - x_1(0)\| \ \ (t \geq 0)$$

for arbitrary solutions $x_1(t), x_2(t)$ of (1.1).

Here $\|.\|$ means the Euclidean norm in $\mathbf{C}^n$, again. It is assumed that there are constants q and ν such that

$$\|A(t) - A(s)\| \leq q\, |t - s| \ (t, s \geq 0), \tag{1.2}$$

and

$$\|F(h_1, t) - F(h_2, t)\| \leq \nu \|h_1 - h_2\|$$
$$(h_1, h_2 \in \mathbf{C}^n;\ t \geq 0). \tag{1.3}$$

Furthermore, it is supposed that

$$v = \sup_{t \geq 0} g(A(t)) < \infty \text{ and } \rho \equiv -\sup_{t \geq 0} \alpha(A(t)) > 0, \tag{1.4}$$

and

$$\sum_{j=0}^{n-1} \frac{v^j}{\sqrt{j!}} \left[\frac{q(j+1)}{\rho^{j+2}} + \frac{\nu}{\rho^{j+1}}\right] < 1. \tag{1.5}$$

Theorem 12.1.2 *Let conditions (1.2)-(1.5) hold. Then system (1.1) is globally exponentially orbitally stable*

Proof: Let x_1 and x_2 be arbitrary solutions of (1.1):

$$\dot{x_k} = A(t)x_k + F\,(x_k, t)\ (k = 1, 2;\ t \geq 0). \tag{1.6}$$

Hence, taking the difference of this equations with $k = 1$ and $k = 2$, we get

$$\dot{x} = A(t)x + \Phi(x, t) \tag{1.7}$$

where

$$x = x_2 - x_1,\ \ \Phi(x, t) = F\,(x_1 + x, t) - F\,(x_1, t).$$

According to (1.3)

$$\|\Phi(x, t)\| \leq q\|x\|\ (x \in \mathbf{C}^n,\ t \geq 0).$$

Now applying to this equation Corollary 7.2.2, we arrive at the required result. □

12.2 One Contour Systems

Consider the equation

$$\dot{y} = Ay + b\ f(s, t)\ (s = cy,\ t \geq 0), \tag{2.1}$$

where A is a real constant Hurwitz $n \times n$-matrix, b is a real column, c is a real row, f maps $\mathbf{R}^1 \times [0, \infty)$ into $\mathbf{R}^1$.

Let us assume that $f(s, t)$ is continuous in t and has the Lipschitz property in the first argument:

$$|f(s_1, t) - f(s_2, t)| \leq \nu|s_1 - s_2| \text{ for all } s_1, s_2 \in \mathbf{R}^1 \text{ and } t \geq 0. \tag{2.2}$$

Certainly, under this condition, equation (2.1) has solutions for all $t \geq 0$.

Let $W(\lambda)$ be the transfer function of the linear part of system (2.1):

$$W(\lambda) = c(\lambda I - A)^{-1}b = P^{-1}(\lambda)L(\lambda)\ (\lambda \in \mathbf{C}),$$

where $P(\lambda)$ and $L(\lambda)$ are polynomials, again.

In addition,

$$P(\lambda) = \lambda^n + a_1\lambda^{n-1} + ... + a_n$$

is a Hurwitz polynomial and $n \equiv \deg P(\lambda) > m \equiv deg\ L(\lambda)$. Besides, $P(\lambda)$ and $L(\lambda)$ have no common roots.

Let

$$K(t) \equiv \frac{1}{2\pi}\int_{-\infty}^{\infty} exp[i\omega t]W(i\omega)\ d\omega\ \ (t \geq 0)$$

be the impulse function of the linear part of (2.1).

Theorem 12.2.1 *Let the conditions (2.2) and*

$$K(t) \geq 0 \text{ for all } t \geq 0 \tag{2.3}$$

be fulfilled. Then for the global orbital exponential stability of (2.1) it is sufficient that the polynomial $P(\lambda) - \nu L(\lambda)$ *be Hurwitzian.*

This theorem is proved in the next section.

Under (2.3), thanks to Lemma 6.1.3, the polynomial $P(\lambda) - qL(\lambda)$ is a Hurwitz one, provided

$$P(0) > \nu L(0). \tag{2.4}$$

This relation and Theorem 12.2.1 imply

Corollary 12.2.2 *Let conditions (2.2)-(2.4) be fulfilled. Then equation (2.1) is globally exponentially orbitally stable.*

Combining this corollary with Lemma 6.1.5, we get

Corollary 12.2.3 *Let all roots of* $P(\lambda)$ *belong to a real segment* $[a, b]$ $(b \leq 0)$. *In addition, let*

$$L^{(k)}(\lambda) \geq 0 \ (a \leq \lambda \leq b;\ k = 0, ..., degL(\lambda)).$$

Then equation (2.1) under (2.2) is is globally exponentially orbitally stable, provided inequality (2.4) is fulfilled. In particular, let $L(\lambda) \equiv 1$, *and let all the roots of* $P(\lambda)$ *be real. Then equation (2.1) is globally orbitally stable if* $P(0) = a_n > \nu$.

With a real Hurwitz polynomial

$$P_3(\lambda) = \lambda^3 + a_1\lambda^2 + a_2\lambda + a_3$$

let us consider the system

$$P_3(D)x = f(x, t) \ \ (D = d/dt). \tag{2.5}$$

Thanks to Lemma 6.1.7, we get

Corollary 12.2.4 *Let* $P_3(\lambda)$ *have a pair of complex conjugate roots:* $-\gamma \pm i\omega$, *and a real root* $-r$ $(r, \gamma, \omega > 0)$. *In addition, let*

$$\gamma > r.$$

Then equation (2.5) is globally orbitally exponentially stable, provided the relations (2.2) and $a_3 > \nu$ *hold.*

12.3 Proof of Theorem 12.2.1

Let y_1 and y_2 be arbitrary solutions of (2.1):

$$\dot{y_k} = Ay_k + b\, f(s_k, t)\ (s_k = cy_k;\ k = 1, 2;\ t \geq 0). \tag{3.1}$$

Hence, taking the difference of this equations with $k = 1$ and $k = 2$, we get

$$\dot{x} = Ax + b\, \Phi(cx, t) \tag{3.2}$$

where

$$x = y_2 - y_1,\ \ \Phi(cx, t) = f(cs_2, t) - f(cs_1, t) = f(cy_1 + cx, t) - f(cy_1, t).$$

According to (2.2)

$$|\Phi(s, t)| \leq \nu|s|\ (s \in \mathbf{R},\ t \geq 0). \tag{3.3}$$

Equation (3.2) is equivalent to following one:

$$x(t) = exp[At]x(0) + \int_0^t exp[A(t-\tau)]b\Phi(cx(\tau), \tau)d\tau.$$

Multiplying this equation by c and taking into account that $c\exp[At]b = K(t)$, we have

$$s(t) = h(t) + \int_0^t K(t-\tau)\Phi(s(\tau), \tau)d\tau, \tag{3.4}$$

where

$$s(t) = cx(t) \text{ and } h(t) = cexp[At]x(0). \tag{3.5}$$

Due to (3.3), we arrive at the inequality

$$|s(t)| \leq |h(t)| + \int_0^t K(t-\tau)\nu|s(\tau)|d\tau.$$

Since A is a Hurwitz matrix, there are constants $\alpha_0, c_1 > 0$ such that

$$|h(t)| \leq c_1 e^{-\alpha_0 t}\ (t \geq 0).$$

In addition,

$$c_1 \leq a_1\|x(0)\|_{R^n}\ (a_1 = const) \tag{3.6}$$

Consequently,

$$|s(t)| \leq c_1 e^{-\alpha_0 t} + \int_0^t K(t-\tau)\nu|s(\tau)|d\tau.$$

From the Lemma 1.7.1 it follows that

$$|s(t)| \leq \eta(t), \tag{3.7}$$

where η is the solution of the equation

$$\eta(t) = c_1 e^{-\alpha_0 t} + \int_0^t K(t-\tau)\nu\eta(\tau)d\tau.$$

We will solve this equation by the Laplace transformation. Since the transform of the convolution is equal to the product of the transforms, after straightforward calculations and taking into account that $P^{-1}(\lambda)L(\lambda)$ is the Laplace transform of $K(t)$, we get the equation

$$\overline{\eta}(\lambda) = \frac{c_1}{\lambda + \alpha_0} + P^{-1}(\lambda)L(\lambda)\nu\overline{\eta}(\lambda),$$

where $\overline{\eta}(\lambda)$ is the Laplace transform of $\eta(t)$. Thus,

$$\eta(t) = c_1 \frac{1}{2\pi i} \int_{-i\infty}^{i\infty} \exp[\lambda t](P(\lambda) -$$

$$\nu L(\lambda))^{-1} P(\lambda)(\lambda + \alpha_0)^{-1} d\lambda.$$

Since $P(\lambda) - \nu L(\lambda)$ is a Hurwitz polynomial, all the poles of the function under the integral lie in the open left half-plane. So, thanks to the residue theorem there is an $\epsilon > 0$, such that

$$\eta(t) \leq c_1 c_2 e^{-\epsilon t} \ (c_2 = const, \ t \geq 0).$$

Now the relations (3.6) and (3.7) yield the required result. □

12.4 Existence and Stability of Forced Oscillations

Theorem 12.4.1 *Under conditions (2.2)-(2.4), let $f(s,t)$ be T-periodic in t:*

$$f(s,t) = f(s, t+T) \text{ for all } s \in \mathbf{R}^1 \text{ and } t \geq 0. \tag{4.1}$$

Then system (2.1) is globally orbitally exponentially stable and has a T-periodic solution.

Proof: Since f is continuous, from (2.2) it follows that

$$|f(s,t)| \leq \nu|s| + l \ (s \in \mathbf{R}; \ l = const > 0). \tag{4.2}$$

Then due to the well-known Theorem 26.1 (M. Krasnosel'skii et al, 1989), under the condition

$$\nu \max_{k=1,2,\ldots} |W(ikT)| < 1 \tag{4.3}$$

equation (2.1) has at least one periodic solution. As it is proved in Section 6.1, condition (2.3) implies

$$\max_{\omega \in \mathbf{R}} |W(i\omega)| = W(0).$$

So conditions (2.3) and (2.4) imply (4.3). Thus the existence of the periodic solution is proved. The global exponential stability follow from Theorem 12.2.1. □

Clearly, the periodic solution is nontrivial if $\sup_t |f(0,t)| > 0$.

12.5 Examples

Example 12.5.1 *Consider the second order equation*

$$\frac{d^2x}{dt^2} + a_1 \frac{dx}{dt} + a_2 x = b_1 f(x,t) + \frac{df(x,t)}{dt}$$

$$(x = x(t),\ a_1, a_2, b_1 = const > 0). \tag{5.1}$$

Let the polynomial $P(\lambda) = \lambda^2 + a_1\lambda + a_2$ have real roots $\lambda_1 \leq \lambda_2 < 0$. Then under the condition

$$b_1 + \lambda_1 \geq 0 \tag{5.2}$$

we have

$$L(\lambda) = \lambda + b_1 \geq 0,\ \frac{d}{d\lambda} L(\lambda) = 1\ (\lambda_1 \leq \lambda \leq \lambda_2).$$

Thus, if the conditions (5.2) and

$$P(0) = a_2 > \nu L(0) = \nu b_1 \tag{5.3}$$

hold, then by Corollary 12.2.3, the equation (5.1) under (2.2) is globally exponentially orbitally stable. If, in addition, f is periodic in t, then due to Theorem 12.4.1, equation (5.1) has a periodic solution.

Example 12.5.2 *Consider the equation*

$$\frac{d^3x}{dt^3} + a_1 \frac{d^2x}{dt^2} + a_2 \frac{dx}{dt} + a_3 x = b_1 f(x,t) + \frac{df(x,t)}{dt}$$

$$(a_1, a_2, a_3, b_1 = const > 0). \tag{5.4}$$

Let the polynomial

$$P(\lambda) = \lambda^3 + a_1\lambda^2 + a_2\lambda + a_3$$

have real roots $\lambda_1 \leq \lambda_2 \leq \lambda_3 < 0$. Then under condition (5.2), relations

$$L(\lambda) = \lambda + b_1 \geq 0,\ \frac{d}{d\lambda} L(\lambda) = 1\ \ (\lambda_1 \leq \lambda \leq \lambda_3)$$

hold. That is, if

$$P(0) - \nu L(0) = a_3 - \nu b_1 > 0 \tag{5.5}$$

then by Corollary 12.2.3, equation (5.4) under (2.1) is globally exponentially orbitally stable. If, in addition, f is periodic in t, then due to Theorem 12.4.1 under (2.2), (5.2) and (5.5), equation (5.4) has a periodic solution.

13. Existence of Steady States. Positive and Nontrivial Steady States

13.1 Systems of Semilinear Equations

Let us consider in $\mathbf{C}^n$ the nonlinear equation

$$Ax = F(x), \tag{1.1}$$

where A is an invertible matrix, F continuously maps $\Omega(r)$ into $\mathbf{C}^n$ for a positive $r \leq \infty$. Recall that $\|.\|$ is the Euclidean norm and $\Omega(r) = \{x \in \mathbf{C}^n : \|x\| \leq r\}$.

Assume that there are positive constants q and l, such that

$$\|F(h)\| \leq q\|h\| + l \ \ (h \in \Omega(r)). \tag{1.2}$$

Lemma 13.1.1 *Under condition (1.2), let*

$$\|A^{-1}\|(qr + l) \leq r. \tag{1.3}$$

Then equation (1.1) has at least one solution $x \in \Omega(r)$, satisfying the inequality

$$\|x\| \leq \frac{\|A^{-1}\|l}{1 - q\|A^{-1}\|}. \tag{1.4}$$

Proof: Set

$$\Psi(y) = A^{-1}F(y) \ \ (y \in \mathbf{C}^n).$$

Hence,

$$\|\Psi(y)\| \leq \|A^{-1}\|(q\|y\| + l) \leq \|A^{-1}\|(qr + l) \leq r \ (y \in \Omega(r)). \tag{1.5}$$

So due to the Brouwer Fixed Point Theorem, equation (1.1) has a solution. Moreover, due to (1.3),

$$\|A^{-1}\|q < 1.$$

Now, using (1.5), we easily get (1.4). □

Put

$$R(A) = \sum_{k=0}^{n-1} \frac{g^k(A)}{d_0^{k+1}(A)\sqrt{k!}}$$

where $g(A)$ is defined in Section 1.5, $d_0(A)$ is the lower spectral radius. That is, $d_0(A)$ is the minimum of the absolute values of the eigenvalues $\lambda_1(A), ..., \lambda_n(A)$ of A:

$$d_0(A) := \min_{k=1,\ldots,n} |\lambda_k(A)|.$$

Due to Lemma 1.5.6,

$$\|A^{-1}\| \leq R(A).$$

Now the previous lemma implies

Theorem 13.1.2 *Under condition (1.2), let*

$$R(A)(qr + l) \leq r.$$

Then equation (1.1) has at least one solution $x \in \Omega(r)$, satisfying the inequality

$$\|x\| \leq \frac{R(A)\, l}{1 - qR(A)}.$$

13.2 Fully Nonlinear Systems

Consider the coupled system

$$\sum_{k=1}^{n} a_{jk}(x)x_k = f_j$$

$$(j = 1, ..., n;\ x = (x_j)_{j=1}^n \in \mathbf{C}^n\ ), \tag{2.1}$$

where

$$a_{jk} : \Omega(r) \to \mathbf{C}\ \ (j, k = 1, ..., n)$$

are continuous functions and $f = (f_j) \in \mathbf{C}^n$ is given. We can write out system (2.1) in the form

$$A(x)x = f \tag{2.2}$$

with the matrix

$$A(z) = (a_{jk}(z))_{j,k=1}^n\ \ (z \in \Omega(r)).$$

Theorem 13.2.1 *Let*

$$\inf_{z\in\Omega(r)} d_0(A(z)) \equiv \inf_{z\in\Omega(r)} \min_k |\lambda_k(A(z))| > 0$$

and

$$\theta_r := \sup_{z\in\Omega(r)} \sum_{k=0}^{n-1} \frac{g^k(A(z))}{\sqrt{k!}\, d_0^{k+1}(A(z))} \leq \frac{r}{\|f\|}. \tag{2.3}$$

Then system (2.1) has at least one solution $x \in \Omega(r)$, satisfying the estimate

$$\|x\| \leq \theta_r \|f\|. \tag{2.4}$$

Proof: Tanks to Lemma 1.5.6,

$$\|A^{-1}(z)\| \le \theta_r \ \ (z \in \Omega(r)).$$

Rewrite (2.2) as

$$x = \Psi(x) \equiv A^{-1}(x)f. \tag{2.5}$$

Due to (2.3)

$$\|\Psi(z)\| \le \theta_r \|f\| \le r \ \ (z \in \Omega(r)).$$

So Ψ maps $\Omega(r)$ into itself. Now the required result is due to the Brouwer Fixed Point theorem. □

Corollary 13.2.2 *Let matrix $A(z)$ be normal:*

$$A^*(z)A(z) = A(z)A^*(z) \ \ (z \in \Omega(r)).$$

If, in addition,

$$\|f\| \le r \inf_{z \in \Omega(r)} d_0(A(z)), \tag{2.6}$$

then system (2.1) has at least one solution x satisfying the estimate

$$\|x\| \le \frac{\|f\|}{\inf_{z \in \Omega(r)} d_0(A(z))}.$$

Indeed, if $A(z)$ is normal, then $g(A(z)) \equiv 0$ and

$$\theta_r = \frac{1}{\inf_{z \in \Omega(r)} d_0(A(z))}.$$

Corollary 13.2.3 *Let matrix $A(z)$ be upper triangular:*

$$\sum_{k=j}^{n} a_{jk}(x)x_k = f_j \ \ (j = 1, ..., n). \tag{2.7}$$

In addition with the notations

$$\tau(A(z)) := \sum_{1 \le j < k \le n} |a_{jk}(z)|^2,$$

and

$$\tilde{a}(z) := \min_{j=1,\dots,n} |a_{jj}(A(z)|,$$

let

$$\sup_{z \in \Omega(r)} \sum_{k=0}^{n-1} \frac{\tau^k(A(z))}{\sqrt{k!}\tilde{a}^{k+1}(z)} \le \frac{r}{\|f\|}. \tag{2.8}$$

Then system (2.7) has at least one solution $x \in \Omega(r)$.

Indeed, this result is due to Theorem 13.2.1, since the eigenvalues of a triangular matrix are its diagonal entries, and

$$g(A(z)) = \tau(A(z)) \text{ and } d_0(A(z)) = \tilde{a}(z).$$

The similar result is true for lower triangular systems.

Note that according to the relations

$$g(A_0) \le \sqrt{1/2}N(A_0^* - A_0)$$

for any constant matrix A_0 (see Section 1.5), in the general case, $g(A(z))$ can be replaced by the simple calculated quantity

$$v(A(z)) = \sqrt{1/2}N(A^*(z) - A(z)) = [\sum_{k=1}^{n} |a_{jk}(z) - \overline{a}_{kj}(z)|^2 \ ]^{1/2}. \tag{2.9}$$

Example 13.2.4 *Let us consider the system*

$$a_{j1}(x)x_1 + a_{j2}(x)x_2 = f_j \quad (x = (x_1, x_2) \in \mathbf{C}^2), \tag{2.10}$$

where f_j are given numbers, and continuous scalar-valued functions

$$a_{jk} \ (j, k = 1, 2)$$

are defined on

$$\Omega(r) \equiv \{z \in \mathbf{C}^2 : \|z\| \le r\}.$$

Due to (2.9)

$$g^2(A(z)) \le |a_{21}(z) - \overline{a}_{12}(z)|^2.$$

In addition, $\lambda_{1,2}(A(z))$ are the roots of the polynomial

$$y^2 - t(z)y + b(z),$$

where

$$t(z) = Trace\ (A(z)) = a_{11}(z) + a_{22}(z)$$

and

$$b(z) = det\ (A(z)) = a_{11}(z)a_{22}(z) - a_{12}(z)a_{21}(z).$$

Then

$$d_0(A(z)) := \min_{k=1,2} |\lambda_k(A(z))|.$$

Moreover,

$$\theta_r \le \tilde{\theta}_r := \sup_{z \in \tilde{\Omega}(r)} \frac{1}{d_0(A(z))} + \frac{|a_{21}(z) - \overline{a}_{12}(z)|}{d_0^2(A(z))}.$$

If

$$\|f\|\tilde{\theta}_r \le r,$$

then due to Theorem 13.2.1, system (2.10) has a solution $x \in \Omega(r)$.

13.3 Nontrivial Steady States

Consider the system

$$F_j(x) = 0 \quad (j = 1, ..., n;\ x = (x_j)_{j=1}^n \in \mathbf{C}^n\ ), \tag{3.1}$$

where functions $F_j : \Omega(r) \to \mathbf{C}$ admit the representation

$$F_j(x) = \phi_j(x)\,(\sum_{k=1}^{n} a_{jk}(x)x_k - f_j)\ \ (j = 1, ..., n). \tag{3.2}$$

Here $\phi_j : \Omega(r) \to \mathbf{C}$ are functions with the property $\phi_j(0) = 0$, $a_{jk} : \Omega(r) \to \mathbf{C}$ are continuous functions, and f_j $(j, k = 1, ..., n)$ are given numbers.

In the sequel *it is assumed that at least one of the numbers f_j is non-zero.* Then from (3.2) it follows that

$$\|A(x)\|\|x\| \geq \|f\| > 0.$$

for any solution $x \in \Omega(r)$ of (2.2) (if it exists). So x is non-trivial. Obviously, (3.1) has the trivial (zero) solution. Moreover, if (2.1) has a solution $x \in \Omega(r)$, then x is simultaneously a solution of (3.1). Now Theorem 13.2.1 implies

Theorem 13.3.1 *Let condition (2.3) hold. Then system (3.1) under (3.2) has at least two solutions: the trivial solution and a nontrivial one satisfying estimate (2.4).*

In addition, the previous theorem and Corollary 13.2.2 yield

Corollary 13.3.2 *Let matrix $A(z)$ be normal for any $z \in \Omega(r)$ and condition (2.6) hold. Then system (3.1) under (3.2) has at least two solutions: the trivial solution and a nontrivial one belonging to $\Omega(r)$.*

Theorem 13.3.1 and Corollary 13.2.3 imply

Corollary 13.3.3 *Let matrix $A(z)$ be upper triangular for any $z \in \Omega(r)$. Then under conditions (3.2) and (2.8), system (3.1) has at least two solutions: the trivial solution and a nontrivial one belonging to $\Omega(r)$.*

The similar result is valid if $A(z)$ is lower triangular.

Example 13.3.4 *Let us consider the system*

$$\phi_j(x_1, x_2)(a_{j1}(x)x_1 + a_{j2}(x)x_2 - f_j) = 0 \quad (j = 1, 2;\ x = (x_1, x_2) \in \mathbf{C}^2) \tag{3.3}$$

where functions $\phi_j : \Omega(r) \to \mathbf{C}$ have the property $\phi_j(0, 0) = 0$. In addition, f_j, a_{jk} $(j, k = 1, 2)$ are the same as in Example 13.2.4. Assume that at least one of the numbers f_1, f_2 is non-zero. Then due to Theorem 13.3.1, under conditions (2.3) system (3.3). has in $\Omega(r)$ at least two solutions.

13.4 Positive Steady States

Consider the coupled system

$$u_j - \sum_{k=1,\ k\neq j}^{n} a_{jk}(u)u_k = F_j(u) \ \ (j=1,...,n), \tag{4.1}$$

where

$$a_{jk}, F_j : \Omega(r) \to \mathbf{R} \ \ (j \neq k;\ j,k=1,...,n)$$

are continuous functions. For instance, the coupled system

$$\sum_{k=1}^{n} w_{jk}(u)u_k = f_j \ \ (j=1,...,n) \tag{4.2}$$

where f_j are given real numbers, $w_{jk} : \Omega(r) \to \mathbf{R}$ are continuous functions, can be reduced to (4.1) with

$$a_{jk}(u) \equiv -\frac{w_{jk}(u)}{w_{jj}(u)}$$

and

$$F_j(u) \equiv \frac{f_j}{w_j(u)},$$

provided

$$w_{jj}(z) \neq 0 \ (z \in \Omega(r);\ j=1,...,n). \tag{4.3}$$

We can write out system (4.1) in the form (1.1) with

$$A(u) = (a_{jk}(u))_{j,k=1}^{n},\ F(u) = column\ (F_j(u))_{j=1}^{n}.$$

Put

$$c_r(F) = \sup_{z\in\Omega(r)} \|F(z)\|.$$

Let $V_+(z)$ and $V_-(z)$ be the upper triangular, lower triangular parts of matrix $A(z)$, respectively:

$$V_+(z) = \begin{pmatrix} 0 & a_{12}(z) & \dots & a_{1n}(z) \\ 0 & 0 & \dots & a_{2n}(z) \\ . & \dots & . & . \\ 0 & 0 & \dots & 0 \end{pmatrix},\ V_-(z) = \begin{pmatrix} 0 & \dots & 0 & 0 \\ a_{21}(z) & \dots & 0 & 0 \\ . & \dots & . & \\ a_{n1}(z) & \dots & a_{n,n-1}(z) & 0 \end{pmatrix}.$$

Recall that $N(A)$ is the Frobenius norm of a matrix A. So

$$N^2(V_+(z)) = \sum_{j=1}^{n-1}\sum_{k=j+1}^{n} a_{jk}^2(z),\ \ N^2(V_-(z)) = \sum_{j=1}^{n}\sum_{k=2}^{j-1} a_{jk}^2(z).$$

Put

$$\tilde{J}(V_\pm(z)) \equiv \sum_{k=0}^{n-1} \frac{N^k(V_\pm(z))}{\sqrt{k!}}.$$

Theorem 13.4.1 *Let the conditions*

$$\alpha_r \equiv \max\{ \inf_{z\in\Omega_r} (\frac{1}{\tilde{J}(V_-(z))} - \|V_+(z)\|),$$

$$\inf_{z\in\Omega_r} (\frac{1}{\tilde{J}(V_+(z))} - \|V_-(z)\|)\} > 0 \quad (4.4)$$

and

$$c_r(F) < r\alpha_r(\Omega_r) \quad (4.5)$$

hold. Then system (4.1) has at least one solution $u \in \Omega(r)$ satisfying the inequaly

$$\alpha_r(\Omega_r)\ \|u\| \le c_r(F). \quad (4.6)$$

In addition, let

$$a_{jk}(z) \ge 0 \text{ and } F_j(z) \ge v_j$$

$$(z \in \mathbf{R}^n : \|z\| \le \frac{c_r(F)}{\alpha_r(\Omega_r)}\ ;\ j \ne k;\ j,k = 1,...,n) \quad (4.7)$$

for some non-negative constants v_j. Then u is non-negative

For the proof see (Gil', 2003c).

13.5 Systems with Differentiable Entries

Consider the system

$$f_k(y_1,...,y_n) = h_k \in \mathbf{C}^1,\ (k = 1,....,n)$$

where $f_j(x) = f_j(x_1,x_2,...,x_n)$, $f_j(0) = 0$ $(j = 1,...,n)$ are scalar-valued differentiable functions defined and continuous with their derivatives on $\mathbf{C}^n$. Put $F(x) = (f_j(x))_{j=1}^n$ and

$$F'(x) \equiv (\frac{\partial f_i(x)}{\partial x_j})_{i,j=1}^n.$$

That is, $F'(x)$ is the Jacobian matrix. Rewrite the considered system as

$$F(y) = h \in \mathbf{C}^n, \quad (5.1)$$

For a positive number $r \le \infty$ assume that

$$\rho_0(r) \equiv \min_{x\in\Omega(r)} d_0(F'(x)) = \min_{x\in\Omega(r)} \min_k |\lambda_k(F'(x))| > 0, \quad (5.2)$$

and

$$\tilde{g}_0(r) = \max_{x\in\Omega(r)} g(F'(x)) < \infty. \quad (5.3)$$

Finally put

$$p(F,r) \equiv \sum_{k=0}^{n-1} \frac{\tilde{g}_0^k(r)}{\sqrt{k!}\rho_0^{k+1}(r)}.$$

Theorem 13.5.1 *Let $f_j(x) = f_j(x_1, x_2, ..., x_n)$, $f_j(0) = 0$ $(j = 1, ..., n)$ be scalar-valued differentiable functions, defined and continuous with their derivatives in $\Omega(r)$. Assume that conditions (5.2) and (5.3) hold. Then for any $h \in \mathbf{C}^n$ with the property*

$$\|h\| \leq \frac{r}{p(F,r)},$$

there is a solution $y \in \mathbf{C}^n$ of system (5.1) which subordinates to the inequality

$$\|y\| \leq \frac{\|h\|}{p(F,r)}.$$

For the proof of this result see (Gil', 1996).

Appendix A.
Bounds for Eigenvalues of Matrices

Although excellent computer software are now available for eigenvalue computation, analytical results on spectrum inclusion regions for finite matrices are still important, since computers are not very useful, in particular, for analysis of matrices dependent on parameters. Here we present some well-known bounds for eigenvalues.

Let $A = (a_{jk})$ be a real $n \times n$-matrix $(n \geq 2)$ with the nonzero diagonal: $a_{kk} \neq 0$ $(k = 1, ..., n)$. Put

$$P_j = \sum_{k=1,\ k\neq j}^{n} |a_{jk}|.$$

A1. In (Marcus and Minc, 1964, Section 3.2.2) it is shown that the eigenvalues of A lie in the union of the sets

$$\{\lambda \in \mathbf{C} : |\lambda - a_{jj}| \leq P_j\} \quad (j = 1, ..., n).$$

A2. In (Marcus and Minc, 1964, Section 3.2.4), it is shown that the spectrum of A lies in the sets

$$\{\lambda \in \mathbf{C} : |a_{ii} - \lambda||a_{jj} - \lambda| \leq P_j P_i\} \quad (i, j = 1, ..., n).$$

A3. Let V_+, V_- and D be the upper nilpotent part, lower nilpotent part and diagonal of A, respectively:

$$V_+ = \begin{pmatrix} 0 & a_{12} & \dots & a_{1n} \\ 0 & 0 & \dots & a_{2n} \\ . & \dots & . & . \\ 0 & 0 & \dots & 0 \end{pmatrix}, \quad V_- = \begin{pmatrix} 0 & \dots & 0 & 0 \\ a_{21} & \dots & 0 & 0 \\ . & \dots & . & . \\ a_{n1} & \dots & a_{n,n-1} & 0 \end{pmatrix}$$

and

$$D = diag\,(a_{11}, a_{22}, ..., a_{nn}).$$

Put

$$w_n = \sum_{j=0}^{n-1} \gamma_{n,j},$$

where $\gamma_{n,j}$ are defined in Section 1.5. One can replace w_n by

$$\tilde{w}_n = \sum_{j=0}^{n-1} \frac{1}{\sqrt{k!}}.$$

Let $V_+ \neq 0, V_- \neq 0$ and assume that

$$w_n \min \{\frac{\|V_-\|}{N(V_+)}, \frac{\|V_+\|}{N(V_-)}\} \leq 1.$$

Here $\|.\|$ is the Euclidean norm and $N(.)$ is the Frobenius (Hilbert-Schmidt) norm. Put

$$\delta_2(A) = w_n^{1/n} \; min \; \{N^{1-1/n}(V_+)\|V_-\|^{1/n}, N^{1-1/n}(V_-)\|V_+\|^{1/n}\}.$$

Then all the eigenvalues of matrix $A = (a_{jk})_{j,k=1}^n$ lie in the union of the discs

$$\{\lambda \in \mathbf{C} : |\lambda - a_{kk}| \leq \delta_2(A)\}, \; k = 1, ..., n. \quad (1)$$

For the proof see (Gil', 2003, Theorem 4.5.1). Thus A is a Hurwitz matrix, provided

$$a_{kk} + \delta_2(A) < 0, \; k = 1, ..., n.$$

A4. Recall that $g(A)$ is defined in Section 1.5. Denote by $z(\nu)$ the unique positive root of the equation

$$z^n(A) = \nu(A) \sum_{k=0}^{n-1} g^k(A)\gamma_{n,k} z^{n-k-1},$$

where

$$\nu(A) = \min\{\|V_-\|, \; \|V_+\|\}.$$

Here one can replace $\gamma_{n,k}$ by $\frac{1}{\sqrt{k!}}$. Then for any $k = 1, ..., n$, there is an eigenvalue μ_0 of A, such that

$$|\mu_0 - a_{kk}| \leq z(\nu). \quad (2)$$

Moreover, the following inequalities are true:

$$\alpha(A) \geq \max_{k=1,\ldots,n} a_{kk} - z(\nu). \quad (3)$$

and $z(\nu) \leq \Delta(A)$, where

$$\Delta(A) = \begin{cases} \nu(A)w_n & \text{if } \nu(A)w_n \geq g(A), \\ g^{1-1/n}(A)[\nu(A)w_n]^{1/n} & \text{if } \nu(A)w_n \leq g(A) \end{cases}.$$

For the proofs see (Gil', 2003, Section 4.6).

A5. Let us point bounds of the other type. Denote

$$q_{up} = \max_{j=2,\ldots,n} \sum_{k=1}^{j-1} |a_{jk}|, \; q_{low} = \max_{j=1,\ldots,n-1} \sum_{j+1}^{n} |a_{jk}|,$$

$$\tilde{v}_k := \max_{j=1,\ldots,k-1} |a_{jk}| \;\; (k = 2, ..., n),$$

$$\tilde{w}_k := \max_{j=k+1,\ldots,n} |a_{jk}| \;\; (k = 1, ..., n-1),$$

$$m_{up}(A) := \prod_{k=2}^{n} (1 + \frac{\tilde{v}_k}{|a_{kk}|}) \text{ and } m_{low}(A) := \prod_{k=1}^{n-1} (1 + \frac{\tilde{w}_k}{|a_{kk}|}).$$

Without losing of generality, assume that

$$min\; \{q_{low} m_{up}(A),\; q_{up} m_{low}(A)\} \le 1. \tag{4}$$

Then all the eigenvalues of A lie in the union of the discs

$$\{\lambda \in \mathbf{C} : |\lambda - a_{kk}| \le \delta_\infty(A)\}, \; k = 1, ..., n,$$

where

$$\delta_\infty(A) := \sqrt[n]{min\{q_{low} m_{up}(A),\; q_{up} m_{low}(A)\}}.$$

For the proof see (Gil', 2003, Theorem 4.7.1).

Under (4), let $\max_j a_{jj} + \delta_\infty(A) < 0$. Then A is a Hurwitz matrix.

Appendix B. Positivity of the Green Function

Equations with Positive "Characteristic" Roots

In this appendix the Cauchy problem for a higher order linear nonautonomous ODE on the positive half-line is considered. Explicit conditions for the positivity of the Green function and its derivatives are derived. Moreover, lower estimates for the Green function are established.

Let $a_k(t)$ $(t \geq 0;\ k = 1, ..., n)$ be real continuous scalar-valued functions defined and bounded on $[0, \infty)$, and $a_0 \equiv 1$. Consider the equation

$$\sum_{k=0}^{n} (-1)^k a_k(t) \frac{d^{n-k} u(t)}{dt^{n-k}} = 0 \ \ (t > 0). \qquad (1.1)$$

A solution of (1.1) is a function $x(.)$ defined on $[0, \infty)$, having continuous derivatives up to the n-th order. In addition, $x(.)$ satisfies (1.1) for all $t > 0$ and the corresponding initial conditions. A scalar valued function $W(t, \tau)$ defined for $t \geq \tau \geq 0$ is *the Green function to equation (1.1)* if it satisfies that equation for $t > \tau$ and the initial conditions

$$\lim_{t \downarrow \tau} \frac{\partial^k W(t,\tau)}{\partial t^k} = 0 \ (k = 0, ..., n-2); \ \lim_{t \downarrow \tau} \frac{\partial^{n-1} W(t,\tau)}{\partial t^{n-1}} = 1.$$

Put

$$c_{2k} = \sup_{t \geq 0} a_{2k}(t), c_{2k-1} = \inf_{t \geq 0} a_{2k-1}(t) \ \ (k = 1, ..., [n/2]), \qquad (1.2)$$

where $[x]$ is the integer part of $x > 0$.

Theorem 15.1.1 *Let all the roots of the polynomial*

$$Q(z) = \sum_{k=0}^{n} (-1)^k c_k z^{n-k} \ \ (c_0 = 1,\ z \in \mathbf{C})$$

be real and nonnegative. Then the Green function to equation (1.1) and its derivatives up to $(n-1)$-th order are nonnegative. Moreover,

$$\frac{\partial^j W(t,\tau)}{\partial t^j} \geq e^{r_1(t-\tau)} \sum_{k=0}^{j} C_j^k \frac{r_1^{j-k}(t-\tau)^{n-1-k}}{(n-1-k)!} \geq 0$$

$$(j = 0, ..., n-1;\ t > \tau \geq 0), \tag{1.3}$$

where $r_1 \geq 0$ *is the smallest root of* $Q(z)$ *and* $C_j^k = \frac{j!}{(j-k)!k!}$.

Below we prove this theorem and remove the condition $r_1 \geq 0$.

Proof of Theorem 15.1.1

Lemma 15.2.1 *Let all the roots of polynomial* $Q(z)$ *be real and nonnegative. Then a solution* u *of (1.1) with the initial conditions*

$$u^{(j)}(0) = 0, j = 0, ..., n-2;\ u^{(n-1)}(0) = 1 \tag{2.1}$$

satisfies the inequalities

$$u^{(j)}(t) \geq e^{r_1 t} \sum_{k=0}^{j} C_k^j \frac{r_1^{j-k} t^{n-1-k}}{(n-1-k)!} \geq 0 \ \ (j = 0, ..., n-1;\ t > 0).$$

Proof: We have

$$b_k(t) := (-1)^k (c_k - a_k(t)) \geq 0 \ \ (k = 1, ..., n).$$

Rewrite equation (1.1) in the form

$$\sum_{k=0}^{n} (-1)^k c_k \frac{d^{n-k} u}{dt^{n-k}} = \sum_{k=1}^{n} b_k(t) \frac{d^{n-k} u}{dt^{n-k}}. \tag{2.2}$$

Denote

$$G(t) = \frac{1}{2i\pi} \int_C \frac{e^{zt} dz}{Q(z)},$$

where C is a smooth contour surrounding all the zeros of $Q(z)$. That is, G is the Green functions to the autonomous equation

$$\sum_{k=0}^{n} (-1)^k c_k \frac{d^{n-k} w(t)}{dt^{n-k}} = 0. \tag{2.3}$$

Put

$$v(t) \equiv \sum_{k=0}^{n} (-1)^k c_k \frac{d^{n-k} u(t)}{dt^{n-k}}. \tag{2.4}$$

Then thanks to the variation of constants formula,

$$u(t) = w(t) + \int_0^t G(t-s) v(s) ds,$$

where $w(t)$ is a solution of equation (2.3). Since

$$G^{(j)}(t) = \frac{1}{2i\pi} \int_C \frac{z^j e^{zt} dz}{(z - r_1)...(z - r_n)},$$

where $r_1 \le ... \le r_n$ are the roots of $Q(z)$ with their multiplicities, due to Lemma 1.11.2 of the present book, we get

$$G^{(j)}(t) = \frac{1}{(n-1)!} [\frac{d^{n-1} z^j e^{zt}}{dz^{n-1}}]_{z=\theta}$$

with a $\theta \in [r_1, r_n]$. Hence,

$$G^{(j)}(t) = \sum_{k=0}^{j} \frac{j! e^{\theta t} \theta^{j-k} t^{n-1-k}}{(j-k)!(n-1-k)!k!} \ge \sum_{k=0}^{j} \frac{j! e^{r_1 t} r_1^{j-k} t^{n-1-k}}{(j-k)!(n-1-k)!k!} \ge 0. \quad (2.5)$$

According to the initial conditions (2.1), we can write out $w(t) = G(t)$. So

$$u(t) = G(t) + \int_0^t G(t-s)v(s)ds. \quad (2.6)$$

Substitute this relation into (2.2). For $j \le n-2$ we have $G^{(j)}(0) = 0$ and

$$\frac{d^j}{dt^j} \int_0^t G(t-s)v(s)ds = \frac{d}{dt} \int_0^t G^{(j-1)}(t-s)v(s)ds =$$

$$\int_0^t G^{(j)}(t-s)v(s)ds \;\; (j = 1, ..., n-1).$$

Hence thanks to (2.2) and (2.4),

$$v(t) = \sum_{k=1}^{n} b_k(t)[G^{(n-k)}(t) + \int_0^t G^{(n-k)}(t-s)v(s)ds] =$$

$$K(t,t) + \int_0^t K(t, t-s)v(s)ds, \quad (2.7)$$

where

$$K(t,\tau) = \sum_{k=1}^{n} b_k(t) G^{(n-k)}(\tau) \;\; (t, \tau \ge 0).$$

According to (2.5), $K(t,\tau) \ge 0 \;\; (t, \tau \ge 0)$. Put $h(t) = K(t,t)$. Let V be the Volterra operator with the kernel $K(t, t-s)$. Then thanks to (2.7) and the Neumann series,

$$v(t) = h(t) + \sum_{k=1}^{\infty} (V^k h)(t) \ge h(t) \ge 0.$$

Hence (2.6) yields,

$$u^{(j)}(t) = G^{(j)}(t) + \int_0^t G^{(j)}(t-s)v(s)ds \geq$$

$$G^{(j)}(t) + \int_0^t G^{(j)}(t-s)K(s,s)ds \geq G^{(j)}(t) \ \ (j = 1, ..., n-1). \qquad (2.8)$$

This and (2.5) prove the lemma. □

Proof of Theorem 15.1.1: For a $\tau > 0$, take the initial conditions

$$u^{(j)}(\tau) = 0, j = 0, ..., n-2; \ u^{(n-1)}(\tau) = 1.$$

Then the corresponding solution $u(t)$ to (1.1) is equal to $W(t,\tau)$. Repeat the arguments of the proof of Lemma 15.2.1. Then instead of (2.8) we have

$$\frac{\partial^j W(t,\tau)}{\partial t^j} \geq G^{(j)}(t-\tau) + \int_\tau^t G^{(j)}(t-\tau-s)K(s,s-\tau)ds \geq G^{(j)}(t-\tau).$$

According to (2.5) this proves the theorem. □

Equations with Arbitrary Real "Characteristic" Roots

In this section we do not assume that the roots of $Q(z)$ are nonnegative. Namely, let $p_k(t)$ $(k = 1, ..., n)$ be real continuous functions bounded on $[0,\infty)$, and $p_0 \equiv 1$. Consider the equation

$$\sum_{k=0}^n p_k(t)\frac{d^{n-k}x}{dt^{n-k}} = 0 \ \ (t > 0). \qquad (3.1)$$

Let the polynomial

$$P(t,z) = \sum_{k=0}^n p_{n-k}(t)z^k \ \ (z \in \mathbf{C})$$

have the purely real roots $\rho_k(t)$ $(k = 1, ..., n)$ with the property

$$\rho_k(t) \geq -\mu \ \ (t \geq 0; \ k = 1, ..., n) \qquad (3.2)$$

with some $\mu > 0$. Put $x(t) = e^{-\mu t}u(t)$ in (3.1). Then

$$0 = e^{\mu t}\sum_{k=0}^n p_k(t)\frac{d^{n-k}e^{-\mu t}u}{dt^{n-k}} = \sum_{k=0}^n p_k(t)(\frac{d}{dt} - \mu)^{n-k}u.$$

That is, equation (3.1) is reduced to the equation

$$P(t, \frac{d}{dt} - \mu)u \equiv \sum_{k=0}^{n} p_k(t)(\frac{d}{dt} - \mu)^{n-k} u = 0. \tag{3.3}$$

But

$$P(t, z-\mu) = \sum_{k=0}^{n} p_k(t)(z-\mu)^{n-k} = \sum_{k=0}^{n} p_k(t) \sum_{j=0}^{n-k} C_{n-k}^{j} (-\mu)^j z^{n-k-j} =$$

$$\sum_{k=0}^{n} p_k(t) \sum_{m=k}^{n} C_{n-k}^{m-k} (-\mu)^{m-k} z^{n-m} = \sum_{m=0}^{n} z^{n-m} \sum_{k=0}^{m} p_k(t) C_{n-k}^{m-k} (-\mu)^{k-m}.$$

That is,

$$P(t, z-\mu) = \sum_{m=0}^{n} (-1)^m q_m(t) z^{n-m},$$

where

$$q_m(t) = \sum_{k=0}^{m} p_k(t) C_{n-k}^{m-k} (-1)^k \mu^{m-k} \quad (m = 1, ..., n), \; q_0 \equiv 1. \tag{3.4}$$

Take into account that

$$P(t, z-\mu) = \prod_{k=1}^{n} (z - \rho_k(t) - \mu) = \prod_{k=1}^{n} (z - \tilde{\rho}_k(t)),$$

where according to (3.2), $\tilde{\rho}_k(t) \equiv \rho_k(t) + \mu \geq 0$. Hence it follows that $q_m(t)$ are nonnegative and we can apply Theorem 15.1.1 to equation (3.3). To this end put

$$d_0 = 1, d_{2k} = \sup_{t \geq 0} q_{2k}(t) \text{ and } d_{2k-1} = \inf_{t \geq 0} q_{2k-1}(t) \; (k = 1, ..., [n/2]), \tag{3.5}$$

where $q_k(t)$ are defined by (3.4). Due to Theorem 15.1.1 and the substitution $x(t) = e^{-\mu t} u(t)$, we get

Theorem 15.3.1 *Under (3.2), let all the roots of the polynomial*

$$\tilde{Q}(z) := \sum_{k=0}^{n} (-1)^k d_k z^{n-k}$$

be real and nonnegative. Then the Green function $\tilde{W}(t, \tau)$ to equation (3.1) is nonnegative and

$$\frac{\partial^j e^{\mu(t-\tau)} \tilde{W}(t,\tau)}{\partial t^j} \geq e^{\tilde{r}_1(t-\tau)} \sum_{k=0}^{j} C_j^k \frac{\tilde{r}_1^{j-k} (t-\tau)^{n-1-k}}{(n-1-k)!} \geq 0$$

$$(j = 0, ..., n-1;\ t > \tau \geq 0),$$

where $\tilde{r}_1 \geq 0$ *is the smallest root of* $\tilde{Q}(z)$. *In particular,*

$$\tilde{W}(t,\tau) \geq e^{(-\mu+\tilde{r}_1)(t-\tau)} \frac{(t-\tau)^{n-1}}{(n-1)!} \quad (t > \tau \geq 0). \tag{3.6}$$

Example 15.3.2 *Let us consider the equation*

$$\frac{d^2x}{dt^2} + p_1(t)\frac{dx}{dt} + p_2(t)x = 0 \quad (t > 0). \tag{3.7}$$

Assume that $p_1(t), p_2(t) \geq 0$ and $p_1^2(t) > 4p_2(t) \quad (t \geq 0)$. Put

$$p_1^+ = \sup_{t \geq 0} p_1(t).$$

Under consideration $\rho_1(t) + \rho_2(t) = -p_1(t)$. So we can take $\mu = p_1^+$. Hence,

$$q_1(t) = 2p_1^+ - p_1(t),\ q_2(t) = p_1^+ \,(p_1^+ - p_1(t)) + p_2(t)$$

and

$$d_1 = \inf_t q_1(t) = p_1^+,\ d_2 = \sup_t q_2(t).$$

If, in addition, $(p_1^+)^2 > 4d_2$, then due to Theorem 15.3.1, the Green function $\tilde{W}(t,\tau)$ to equation (3.7) is nonnegative and inequality (3.6) is valid with $n = 2$ and

$$-\mu + \tilde{r}_1 = -p_1^+/2 - \sqrt{(p_1^+)^2/4 - d_2}.$$

Third Order Equations with Nonreal Unstable "Characteristic" Roots

Let $a_k(t)$ $(t \geq 0;\ k = 1,2,3)$ be real continuous scalar-valued functions bounded on $[0,\infty)$ and $a_0 \equiv 1$. Consider the equation

$$\sum_{k=0}^{3} (-1)^k a_k(t) \frac{d^{3-k}u(t)}{dt^{3-k}} = 0 \quad (t > 0). \tag{4.1}$$

The function $W(t,\tau)$ defined for $t \geq \tau \geq 0$ is *the Green function to equation (4.1)* if it satisfies (4.1) for $t > \tau$ and the conditions

$$\lim_{t \downarrow \tau} \frac{\partial^k W(t,\tau)}{\partial t^k} = 0 \ (k = 0,1);\ \lim_{t \downarrow \tau} \frac{\partial^2 W(t,\tau)}{\partial t^2} = 1.$$

Put

$$c_1 = \inf_{t \geq 0} a_1(t), c_2 = \sup_{t \geq 0} a_2(t), c_3 = \inf_{t \geq 0} a_3(t). \tag{4.2}$$

Let the polynomial

$$Q_3(z) = \sum_{k=0}^{3}(-1)^k c_k z^{3-k} \quad (c_0 = 1,\ z \in \mathbf{C})$$

have a pair of complex conjugate roots: $\gamma \pm i\omega$ $(\gamma, \omega > 0)$, and a positive root z_0. In addition, let

$$z_0 > \gamma + \omega. \tag{4.3}$$

Denote

$$K(t) := c_0[e^{z_0 t} - e^{\gamma t}(cos\ (\omega t) + b_0\ sin\ (\omega t))],$$

where

$$b_0 := \frac{z_0 - \gamma}{\omega} \text{ and } c_0 := \frac{1}{(z_0 - \gamma)^2 + \omega^2}.$$

Below we prove that $K(t)$ and its first and second derivatives are nonnegative on the half-line.

Theorem 15.4.1 *Let polynomial $Q_3(z)$ have a pair of complex conjugate roots: $\gamma \pm i\omega$ $(\gamma, \omega > 0)$, and a positive root z_0 satisfying condition (4.3). Then the Green function $W(t, \tau)$ to equation (4.1) and its first, and second derivatives are nonnegative. Moreover,*

$$\frac{\partial^k W(t,\tau)}{\partial t^k} \geq \frac{\partial^k K(t-\tau)}{\partial t^k} \quad (k = 0, 1, 2;\ t > \tau \geq 0). \tag{4.4}$$

To prove this theorem we need the following

Lemma 15.4.2 *Let $Q_3(\lambda)$ have a pair of complex conjugate roots: $\gamma \pm i\omega$ $(\gamma, \omega > 0)$, and a positive root z_0. In addition, let condition (4.3) hold. Then*

$$K^{(j)}(t) \geq 0 \ \ (j = 0, 1, 2;\ t \geq 0).$$

Proof: Put $b_1 = z_0 - \gamma$. So $b_1 = \omega b_0$. Since

$$K(t) = c_0 e^{\gamma t} f(t) \text{ where } f(t) = e^{b_1 t} - cos(\omega\, t) - b_0\ sin(\omega t),$$

it is enough to check that f and its derivatives are positive. By virtue of the Taylor series

$$e^{b_1 t} = 1 + b_1 t + g(t) \ \ (g(t) > 0).$$

Since $|cos\ s| \leq 1, |sin\ s| \leq s\ (s \geq 0)$, we have

$$f(t) = 1 + b_1 t + g(t) - cos\ \omega t - b_0\ sin\ \omega t \ \ (g(t) > 0).$$

Thus, one can assert that $f \geq 0$ for all $t \geq 0$. Furthermore, due to (4.3), $b_0 \geq 1$ and $b_1 \geq \omega$. Thus

$$\dot{f}(t) = b_1 e^{b_1 t} + \omega sin(\omega t) - b_0 \omega\ cos(\omega t) =$$

$$b_1(1+b_1t+g(t))+\omega sin(\omega t)-\omega b_0\; cos(\omega t)\geq 0.$$

Moreover,

$$\ddot{f}(t)=b_1^2e^{b_1t}+\omega^2cos(\omega t)+$$

$$\omega^2b_0\; sin(\omega t)=b_1^2(1+b_1t+g(t))+\;\omega^2cos(t\omega)+b_0\omega^2sin(t\omega)\geq 0.$$

As claimed. □

Lemma 15.4.3 *Under the conditions of Theorem 4.1, a solution u of (4.1) with the initial conditions*

$$u(0)=\dot{u}(0)=0;\; \ddot{u}(0)=1 \tag{4.5}$$

satisfies the inequalities $u^{(j)}(t)\geq K^{(j)}(t)\;\; (j=0,1,2;\; t>0)$.

Proof: Simple calculations show that $K(t)$ is a solution of the equation

$$Q_3(D)x(t)=0\;\; (D=d/dt), \tag{4.6}$$

with the initial condition $K(0)=\dot{K}(0)=0,\;\ddot{K}(0)=1$. So K is the Green function to (4.6). Furthermore, we have

$$m_k(t):=(-1)^k(c_k-a_k(t))\geq 0\;\; (k=1,2,3).$$

Rewrite equation (4.1) in the form

$$\sum_{k=0}^{3}(-1)^kc_k\frac{d^{3-k}u}{dt^{3-k}}=\sum_{k=1}^{3}m_k(t)\frac{d^{3-k}u}{dt^{3-k}}. \tag{4.7}$$

Put

$$v(t):=\sum_{k=0}^{3}c_k\frac{d^{3-k}u(t)}{dt^{3-k}}. \tag{4.8}$$

Since, K is the Green functions to (4.6), thanks to the Variation of Constants Formula and condition (4.5), we have

$$u(t)=K(t)+\int_0^t K(t-s)v(s)ds. \tag{4.9}$$

But $\dot{K}(0)=K(0)=0$. Consequently,

$$\frac{d^j}{dt^j}\int_0^t K(t-s)v(s)ds=\frac{d}{dt}\int_0^t K^{(j-1)}(t-s)v(s)ds=$$

$$\int_0^t K^{(j)}(t-s)v(s)ds\;\; (j=1,2).$$

Hence thanks to (4.7) and (4.8),

$$v(t) = \sum_{k=1}^{3} m_k(t)[K^{(3-k)}(t) + \int_0^t K^{(3-k)}(t-s)v(s)ds] =$$

$$K_0(t,t) + \int_0^t K_0(t,t-s)v(s)ds, \tag{4.10}$$

where

$$K_0(t,\tau) = \sum_{k=1}^{3} m_k(t) K_0^{(3-k)}(\tau) \ \ (t, \tau \geq 0).$$

According to the previous lemma, $K_0(t,\tau) \geq 0 \quad (t, \tau \geq 0)$. Put $h(t) = K_0(t,t)$. Let V be the Volterra operator with the kernel $K_0(t, t-s)$. Then thanks to (4.10) and the Neumann series,

$$v(t) = h(t) + \sum_{k=1}^{\infty} (V^k h)(t) \geq h(t) \geq 0.$$

Hence (4.9) and the previous lemma yield

$$u^{(j)}(t) = K^{(j)}(t) + \int_0^t K^{(j)}(t-s)v(s)ds \geq$$

$$K^{(j)}(t) \geq 0 \ \ (j = 0, 1, 2), \tag{4.11}$$

as claimed. □

Proof of Theorem 15.4.1: For a $\tau > 0$, take the initial conditions

$$u(\tau) = \dot{u}(\tau) = 0, j = 0, 1; \ \ddot{u}(\tau) = 1.$$

Then the corresponding solution $u(t)$ to (4.1) is equal to $W(t,\tau)$. Repeat the arguments of the proof of Lemma 15.4.3. Then according to (4.11) we have the required result. As claimed. □

Third Order Equations with General Nonreal "Characteristic" Roots

Let $p_k(t)$ $(k = 1, ..., 3)$ be real continuous functions bounded on $[0, \infty)$, and $p_0 \equiv 1$. Consider the equation

$$\sum_{k=0}^{3} p_k(t) \frac{d^{3-k}x}{dt^{3-k}} = 0 \ \ (t > 0). \tag{5.1}$$

Let the polynomial

$$P_3(t,z) = \sum_{k=0}^{3} p_{3-k}(t) z^k \quad (z \in \mathbf{C})$$

have the roots $\rho_k(t)$ $(k = 1, 2, 3)$ with the property

$$Re\ \rho_k(t) \geq -\mu \quad (t \geq 0;\ k = 1, 2, 3) \tag{5.2}$$

with some $\mu > 0$.

Denote

$$q_1(t) = 3\mu - p_1(t),\ q_2(t) = 3\mu^2 - 2\mu p_1(t) + p_2(t)$$

and

$$q_3(t) = \mu^3 - p_1(t)\mu^2 + p_2(t)\mu - p_3(t),$$

and

$$d_0 = 1, d_1 = \inf_{t\geq 0} q_1(t), d_2 = \sup_{t\geq 0} q_2(t), \text{ and } d_3 = \inf_{t\geq 0} q_3(t).$$

Let the polynomial

$$\tilde{Q}_3(z) = \sum_{k=0}^{3} (-1)^k d_k z^{3-k} \quad (d_0 = 1,\ z \in \mathbf{C})$$

have a pair of complex conjugate roots: $\tilde{\gamma} \pm i\tilde{\omega}$ $(\tilde{\gamma}, \tilde{\omega} > 0)$, and a positive root $\tilde{z}_0$. In addition, let

$$\tilde{z}_0 > \tilde{\gamma} + \tilde{\omega}. \tag{5.3}$$

Denote

$$\tilde{b}_0 := \frac{\tilde{z}_0 - \tilde{\gamma}}{\tilde{\omega}} \text{ and } \tilde{c}_0 := \frac{1}{(\tilde{z}_0 - \tilde{\gamma})^2 + \tilde{\omega}^2}.$$

Theorem 15.5.1 *Let the polynomial $\tilde{Q}_3(z)$ have a pair of complex conjugate roots: $\tilde{\gamma} \pm i\tilde{\omega}$ $(\tilde{\gamma}, \tilde{\omega} > 0)$, and a positive root $\tilde{z}_0$ satisfying condition (5.3). Then the Green function $\tilde{W}(t,\tau)$ to equation (5.1) is nonnegative. Moreover, for all $t > \tau \geq 0$, we have*

$$\tilde{W}(t,\tau) \geq \tilde{c}_0[e^{(\tilde{z}_0-\mu)(t-\tau)} - e^{\tilde{\gamma}-\mu)(t-\tau)}(cos\ (\tilde{\omega}\ (t-\tau)) + \tilde{b}_0\ sin(\tilde{\omega}\ (t-\tau)))] \geq 0$$

Proof: Put $x(t) = e^{-\mu t}u(t)$ in (5.1). Then

$$0 = e^{\mu t} \sum_{k=0}^{3} p_k(t) \frac{d^{3-k} e^{-\mu t} u}{dt^{3-k}} = \sum_{k=0}^{3} p_k(t) (\frac{d}{dt} - \mu)^{3-k} u.$$

That is, equation (5.1) is reduced the equation

$$P_3(t, \frac{d}{dt} - \mu)u \equiv \sum_{k=0}^{3} p_k(t)(\frac{d}{dt} - \mu)^{3-k} u = 0. \tag{5.4}$$

But

$$P_3(t, z-\mu) = \sum_{k=0}^{3} p_k(t)(z-\mu)^{3-k} = \sum_{k=0}^{3} p_k(t) \sum_{j=0}^{3-k} C_{3-k}^{j}(-\mu)^j z^{3-k-j} =$$

$$\sum_{k=0}^{3} p_k(t) \sum_{m=k}^{3} C_{3-k}^{m-k}(-\mu)^{m-k} z^{3-m} = \sum_{m=0}^{3} z^{3-m} \sum_{k=0}^{m} p_k(t) C_{3-k}^{m-k}(-\mu)^{k-m}.$$

where $C_n^k = n!/k!(n-k)!$. That is,

$$P_3(t, z-\mu) = \sum_{m=0}^{3} (-1)^m q_m(t) z^{3-m},$$

where

$$q_m(t) = \sum_{k=0}^{m} p_k(t) C_{3-k}^{m-k}(-1)^k \mu^{m-k} \ \ (m = 1, 2, 3), \ q_0 \equiv 1.$$

Take into account that

$$P_3(t, z-\mu) = \prod_{k=1}^{3} (z - \rho_k(t) - \mu),$$

where according to (5.2), *Re* $\rho_k(t) + \mu \geq 0$ and apply Theorem 15.4.1 to equation (5.4). Due to the substitution $x(t) = e^{-\mu t} u(t)$, Theorem 15.4.1 proves the required result. □

Notes

Chapter 1: This book presupposes a knowledge of basic matrix theory, for which there are good introductory texts. The books (Gantmaher, 1967) and (Bellman, 1970) are classical. For more details about the notions presented in Chapter 1 also see (Stewart and Sun, 1990). The stability definitions presented in Section 1.2 are particularly taken from the books (Reissig et al., 1974), (Bellman, 1953), and (Vidyasagar, 1993). Books that provide a broader look at the subjects we cover include the books (Coppel, 1965), (Harris and Miles, 1980), (Khalil, 1992), (Kohan, 1994), (Nijmeijer and van der Schaft, 1990), (Rugh, 1996), (Slotine and Li, 1991), (Sontag, 1990), (Willems, 1971), etc. The material of Section 1.7 is based on Section 3.1 (Daleckii and Krein, 1974). The material of Sections 1.10-1.12 is adapted from (Gil', 1995). Estimates for roots of algebraic equations can be found in the book (Ostrovski, 1973, p. 277).

Chapter 2: The material of Sections 2.1 and 2.2 is based on Sections 3.1 and 3.2 of the book (Daleckii and Krein, 1974). The multiplicative representation for solutions of linear ordinary differential equations is well-known cf. (Dollard and Friedman, 1979), (Gantmaher, 1967). Theorems 2.6.1 and 2.7.1 are proved in (Gil', 1995). About other very interesting results on stability of linear systems see (Blondel, 1993), (Liu and Michel, 1994).

Chapter 3: The results of Sections 3.1 and 3.2 are adapted from the paper (Gil', 1989b). They are based on the freezing method developed in the book (Bylov et al., 1966), and the papers (Izobov, 1974) and (Vinograd, 1983). Theorem 3.3.1 is taken from the paper (Gil', 2004b). A survey of results on uniform exponential stability under the hypothesis that pointwise eigenvalues of the slowly-varying $A(t)$ have negative real parts is in (Ilchmann et al., 1987). Other very interesting results on stability and instability of slowly varying systems can be found in the papers (Skoog and Lau, 1972) and (Desoer, 1969).

Chapter 4: Theorems 4.1.1 is a corollary of the multiplicative representation. Corollary 4.1.2 was derived by Lozinskii, Corollary 4.1.6 is due to Wazewski (see (Izobov, 1974)). Similar result was established by Winter (1946). For additional results see also (Wu, 1984).

Chapter 5: Theorem 5.1.1 is adapted from (Gil' and Ailon, 1999). It was generalized to retarded systems in (Gil', 1994a). Stabilizability of nonlinear systems with leading autonomous parts have been discussed by many authors (see (Tsinias, 1991), (Vidyasagar, 1993), and references given therein). About some other analytical methods of estimating the domain of attraction for differential equations see for instance (Levin, 1994), (Khalil, 1992), (Siljak, 1978). As it was above mentioned the basic method for the stability analysis of autonomous continuous systems is the Lyapunov functions one, cf. (Rajalaksmy and Sivasundaram, 1992), (Lakshmikantham et al., 1989), (Lakshmikantham et al., 1991), etc. About the well-known relevant results see also (Gelig, Leonov and Yakubovich, 1978), (Leonov, 2001), (Yakubovich, 1998), etc.

Chapter 6: As it was above mentioned, in 1966 N. Truchan has showed that the Aizerman's hypothesis is satisfied by systems having linear parts in the form of single loop circuits with up to five stable aperiodic links connected in tandem, cf. (Voronov, 1979). Theorem 6.1.2 includes Truchan's one. It was announced in (Gil', 1983a) and proved in (1983b). The contents of Section 6.3 are based on the papers (Gil', 1985) and (Gil', 1994b). The paper (Gil' and Shargorodsky, 1986) deals with applications of Theorem 6.1.2 to limit cycles and oscillations of autogenerators. The results presented in Sections 6.1-6.3 supplement the well-known absolute stability criteria cf. (Xiao-Xin, 1993), (Harris and Valenca, 1983), (Vidyasgar, 1993), (Krasnosel'skii and Pokrovskii, 1977), etc. The Aizerman type problem for retarded systems was considered in (Gil', 2000a).

Chapter 7: Theorems 7.1.1 and 7.2.1 are adapted from (Gil' and Ailon, 1999). The contents of Sections 7.3 and 7.4 are taken from the papers (Gil', 2004b).

Chapter 8 is based on the papers (Gil', 1989a) and (Gil', 1989b). The contents of Section 7.8 is adapted from the paper (Levin, 1969). About the theory of perturbations of nonlinear systems see for instance (Bellman, 1953, Section 2.5), (Bellman et al., 1985), (Ladde et al., 1977), (Rajalaksmy and Sivasundaram, 1992), etc. About other interesting relevant results see the papers (Aeyels and Peuteman, 1997 and 1998), (Peuteman and Aeyels, 2002), etc.

The material of **Chapter 9** is probably new.

The contents of **Chapter 10** are taken from the paper (Gil', 2004a).

The results of **Chapter 11** are based on the paper (Gil' and Ailon, 1998). In the papers (Gil', 2000b) and (Gil' and Ailon, 2000) the similar result were derived for retarded systems. About the well-known input-to-state stability results see (Sontag, 1990), (Angeli, Sontag and Wang, 2000), (Arcak and Teel, 2002), (Krichman, Sontag and Wang, 2000), (Nešić and Teel, 2001), (Tsinias, 1999). The notion of input-state stability is closely connected with input-output stability, cf. (Vidyasagar, 1993), (Rugh, 1996, Chapter 12), etc.

The material of **Chapter 12** is based on the paper (Gil', 2001).

The results of **Chapter 13** are taken from the papers (Gil', 1996) and (Gil', 2003d).

References

[1] Aeyels, D. and Peuteman, J. (1997). New criteria for exponential and uniform asymptotic stability of nonlinear time-variant differential equations. *Proceedings of IEEE Conference on Decision and Control*, San Diego.

[2] Aeyels, D. and Peuteman, J. (1998). A new asymptotic stability criterion for nonlinear time-variant differential equations. *IEEE Trans. Autom. Control* **43**, No. 7, 968–971.

[3] Aizerman, M.A. (1949). On a problem concerning global stability of dynamical systems, *Uspekhi Matematicheskikh Nauk*, **4(4)**, 187–188. In Russian.

[4] Angeli, D., Sontag, E.D. and Wang, Y. (2000). Further equivalences and semiglobal versions of integral input to state stability. *Dyn. Control* **10**, No.2, 127–149.

[5] Arcak, M. and Teel, A. (2002). Input-to-state stability for a class of Lur'e systems. *Automatica* **38**, No.11, 1945–1949.

[6] Bellman, R.E. (1953). *Stability Theory of Differential Equations.* McGraw-Hill, New York.

[7] Bellman, R.E. (1970). *Introduction to Matrix Analysis.* McGraw-Hill, New York.

[8] Bellman R.E., Bentsmann, J., and Meerkov, S.M. (1985). Stability of fast periodic systems, *IEEE Transactions on Automatic Control*, **30**, No 3, 289–291.

[9] Blondel, V. (1993). *Simulteneous Stabilization of Linear Systems*, Lecture Notes in Control and Information Sciences, Vol. 191, Springer, London.

[10] Bylov, B.F., Grobman, B. M., Nemyckii V. V. and Vinograd R.E. (1966). *The Theory of Lyapunov Exponents.* Nauka, Moscow. In Russian.

[11] Coppel, W.A. (1978). *Dichotomies in Stability Theory.* Lectures Notes in Mathematics 629, Springer-Verlag, New York.

[12] Daleckii, Yu L. and Krein, M. G. (1974). *Stability of Solutions of Differential Equations in Banach Space*, Amer. Math. Soc., Providence, R. I.

[13] Desoer, C.A. (1969). Slowly varying systems $dotx = A(t)x$, *IEEE Transactions on Automatic Control*, **14**, 780–781.

[14] Döetsch, G. (1961). *Anleitung zum Praktischen Gebrauch der Laplace-transformation.* Oldenburg, Munchen.

[15] Dollard, J.D. and Friedman, Ch. N. (1979). *Product Integration with Applications to Differential Equations.* Encyclopedia of Mathematics and its applications; v.10., London, Addison-Wesley Publ. Company.

[16] Gantmaher, F. R. (1967). *Theory of Matrices.* Nauka, Moscow. In Russian

[17] Gelfond, A. O. (1967). *Calculations of Finite Diferences.* Nauka, Moscow. In Russian.

[18] Gelig, A. K, Leonov, G.A. and Yakubovich, V.A. (1978). *The Stability of Nonlinear Systems with a Nonunique Equilibrium State.* Nauka, Moskow. In Russian

[19] Gil', M.I. (1983a). On a class of one-contour systems which are absolutely stable in the Hurwitz angle, *Automation and Remote Control, No. 10*, 70–75.

[20] Gil', M.I. (1983b). On one class of absolutely stable systems, *Soviet Physics Doklady*, **269(6)**, 1324–1327.

[21] Gil', M.I. (1985). On one class of absolutely stable multivariable systems, *Soviet Physics Doklady*, **280(4)**, 811–815.

[22] Gil', M.I. (1989a). Single-loop systems with several nonlinearities satisfying the generalized Aizerman-Calman conjecture, *Soviet Physics Doklady*, **292 (2)**, 1315–1318.

[23] Gil', M.I. (1989b). The freezing method for nonlinear equations. *Differential Eqs.*, **25**, 912–918.

[24] Gil', M.I. (1994a), On absolute stability of differential-delay systems, *IEEE, Trans. Automatic Control*, **39**, N 12, 2481–2484.

[25] Gil', M.I. (1994b). Class of absolutely stable multivariable systems, *International Journal of Systems Sciences* **25(3)**, 613–617.

[26] Gil', M.I. (1995). *Norm Estimations for Operator-valued Functions and Applications.* Marcel Dekker, Inc, New York.

[27] Gil', M.I. (1996). On solvability of nonlinear equations in lattice normed spaces, *Acta Sci. Math. (Szeged)*, **62**, 201–215

[28] Gil', M.I. (1998). *Stability of Finite and Infinite Dimensional Systems*, Kluwer, New York, 1998

[29] Gil', M.I. (2000a), On Aizerman-Myshkis problem for systems with delay. *Automtica*, **36**, 1669–1673

[30] Gil', M.I. (2000b), On input-output stability of nonlinear retarded systems *Robust and Nonlinear Control* **10**, 1337–1344.

[31] Gil', M.I. (2001), Aizerman's problem for orbital stability and forced oscillations, *ZAMM*, **81**, 564–570.

[32] Gil', M.I. (2003a). *Operator Functions and Localization of Spectra*, Lectures Notes in Mathematics, Vol. 1830, Springer Verlag, Berlin.

[33] Gil', M.I. (2003b). Bounds for the spectrum of analytic quasinormal operator pencils in a Hilbert space, *Contemporary Mathematics*, **5**, No 1, 101–118

[34] Gil', M.I. (2003c). Inner bounds for spectra of linear operators, *Proceedings of the American Mathematical Society*, **131**, 3737–3746.

[35] Gil', M.I. (2003d). On positive solutions of nonlinear equations in a Banach space, *Nonlinear Functional Analysis*, **8**, No. 4, 581–593.

[36] Gil', M.I. (2004a). Differential equations with bounded positive Green's functions and generalized Aizerman's hypothesis, *Nonlinear Differential Equations*, **11**, 137–150

[37] Gil', M.I. (2004b). A new stability test for nonlinear nonautonomous systems, *Automatica* (accepted for publication)

[38] Gil', M.I. and A. Ailon. (1998), The input-output version of Aizerman's conjecture, *Robust and Nonlinear Control* **8**, 1219–1226.

[39] Gil', M.I. and A. Ailon (1999). On exponential stabilization of nonlinear nonautonomous systems, *Int. J. Control*, **72**, No. 5, 430–434

[40] Gil', M.I. and A. Ailon (2000). On input-output stability of nonlinear retarded systems *IEEE Transactions, CAS-I* , **47**, No 5, 1337–1344.

[41] Gil', M. I. and Shargorodsky, L.L. (1986). On one criterion of existence of limit cycles, *Soviet Mathematics*, **30**, 12–14.

[42] Godunov, S.K. (1998). *Modern Aspects of Linear Algebra*, Trans. of Math. Monographs, vol 175, AMS, Providenc, R. I.

[43] Gohberg, I. C. and Krein, M. G. (1970) . *Theory and Applications of Volterra Operators in Hilbert Space*, Trans. Mathem. Monographs, vol. 24, Amer. Math. Soc., R. I.

[44] Harris, C.J. and Miles J.F. (1980). *Stability of Linear Systems*, Academic Press, New York.

[45] Harris, C.J. and Valenca, J. (1983). *The Stability of Input-Output Dynamical Systems.* Academic Press, London-New York.

[46] Ilchmann A., Owens, D.H, and Pratcel-Wolters, D. (1987). Sufficient conditions for stability of linear time-varying systems, *Systems and Control Letters*, **9**, 157–163.

[47] Izobov, N. A. (1974). Linear systems of ordinary differential equations. *Itogi Nauki i Tekhniki. Mat. Analis, 12*: 71–146. In Russian.

[48] Khalil, H.K. (1992). *Nonlinear Systems.* MakMillan, New York.

[49] Kohan, J. (1994). *Robust Stability and Convexity*, Lecture Notes in Control and Information Sciences, Vol. 201, Springer, London.

[50] Krasnosel'skii, M. A., Burd, Sh., and Yu. Kolesov. (1970). *Nonlinear Almost Periodic Oscillations*, Nauka, Moscow. In Russian.

[51] Krasnosel'skii, M. A., Lifshits, J., and A. Sobolev. (1989). *Positive Linear Systems. The Method of Positive Operators.* Heldermann Verlag, Berlin.

[52] Krasnosel'skii, M. A. and Pokrovskii, A. (1977). The absent bounded solution principle, *Soviet Math. Doklady*, **233**, 293–296.

[53] Krichman, M., Sontag, E. D. and Wang, Y. (2000). Input-output-to-state stability. *SIAM J. Control Optimization* **39**, No.6, 1874–1928

[54] Ladde, G.S., Lakshmikantham, V. and S. Leela (1977). A new technique in perturbation theory, *Rocky Mountain Journal Math. , 6*, 133–140.

[55] Lakshmikantham, V., Leela, S., and Martynyuk, A.A. (1989). *Stability Analysis of Nonlinear Systems*, Marsel Dekker, Inc, New York.

[56] Lakshmikantham, V., Matrosov, V. M., and Sivasundaram, S. (1991). *Vector Lyapunov Functions and Stability Analysis of Nonlinear Systems*, Kluwer Academic Publishers, Dordrecht, Boston, London.

[57] Leonov G.A. (2001). *Mathematical Problems of Control theory. An introduction.* World Scientific Publ., Series on Stability, Vibration and Control of Systems, Series A. 4. Singapore.

[58] Levin, A. Yu. (1969). Non-oscillations of solutions of the equation $x^{(n)}(t) + a_1(t)x^{(n-1)}(t) + ... + p_n(t)x(t)$, *Russian Mathematical Surveys*, **24(2)**, 43–96.

[59] Levin, A. (1994). Analytical method of estimating the domain of attraction for polynomial differential equations, *IEEE Transactions on Automatic Control*, **39**, No 12, 2471–2476.

[60] Liu, D. and Michel A.N. (1994). *Dynamical Systems with Saturation Nonlinearities: Analysis and Design*, Lecture Notes in Control and Information Sciences, Vol. 195, Springer, London.

[61] Marcus, M. and Minc, H. (1964). *A Survey of Matrix Theory and Matrix Inequalities.* Allyn and Bacon, Boston.

[62] Naredra, K. S. and J. H. Taylor. (1973). *Frequency Domain Criteria for Absolute Stability*, Academic Press, New York and London.

[63] Nešić, D. and Teel, A.R., (2001). Input-to-state stability for nonlinear time-varying systems via averaging. *Math. Control Signals Syst.* **14**, No.3, 257–280.

[64] Nijmeijer, H. and A. van der Schaft (1990). *Nonlinear Dynamical Control Systems.* Springer-Verlag, New York.

[65] Ostrowski, A. M. (1973). *Solution of Equations in Euclidean and Banach spaces.* Ac. Press, New York-London.

[66] Peuteman, J. and Aeyels, D., (2002). Exponential stability of slowly time-varying nonlinear systems. *Math. Control Signals Syst*, **15**, No.3, 202–228.

[67] Rajalaksmy S. and Sivasundaram, S. (1992). Vector Lyapunov functions and the technique in perturbation theory, *Journal of Mathematical Analysis and Applications*, **164**, 560–570.

[68] Reissig, R., Sansone, G. and R. Conti. (1974). *Nonlinear Differential Equations of High Order.* Noordhoff International Publishing, Leiden.

[69] Rodman, L. (1989). *An Introduction to Operator Polynomials*, Operator Theory. Advances and Appl. v. 38, Birkhäuser, Basel.

[70] Rugh, W.J. (1996). *Linear System Theory.* Prentice Hall, Upper Saddle River, New Jersey.

[71] Siljak, D.D. (1978). *Large Scale Dynamic Systems. Stability and Structure*, North-Holland, New York.

[72] Skoog, R.A. and Lau, G.Y. (1972). Instability of slowly varying systems, *IEEE Transactions on Automatic Control*, **17**, 86–92.

[73] Slotine, J.-J. and W. Li. (1991). *Applied Nonlinear Control.* Prentice-Hall. Englewood Cliffs, New Jersey.

[74] Sontag, E.G. (1990). *Mathematical Control Theory: Deterministic Finite Dimensional Systems.* Springer-Verlag, New York.

[75] Stewart, G. W. and Sun Ji-guang (1990). *Matrix Perturbation Theory*, Academic Press, New York.

[76] Tsinias, J. (1991). A theorem on global stabilization of nonlinear systems by linear feedback, *Systems & Control Letters* **17**, 357–362.

[77] Tsinias, J., (1999). Control Lyapunov functions, input-to-state stability and applications to global feedback stabilization for composite systems. *J. Math. Syst. Estim. Control* **7**, No.2, 235–238.

[78] Vidyasagar, M. (1993). *Nonlinear Systems Analysis*, second edition. Prentice-Hall. Englewood Cliffs, New Jersey.

[79] Vinograd R.E. (1957). The inadequace of the method of characteristic exponents for the study of nonlinear differential equations, *Math. Sbornik*, **41**, 431–438. In Russian.

[80] Vinograd, R.E. (1983). An improved estimate in the method of freezing, *Proc. Amer. Soc.* **89 (1)**, 125–129.

[81] Voronov. A. (1979). *Stability, Controllability, Observability*, Nauka, Moscow. In Russian.

[82] Willems, J. C. (1971). *The Analysis of Feedback Systems*, M. I. T. Press, Cambridge, M. A.

[83] Winter, A. (1946). Asymptotic integration constant, *American Journal of Mathematics*, **68**, 125–132.

[84] Wu, M.Y. (1984). Stability of linear time-varying systems, *International Journal of System Sciences*, **15**, 137–150.

[85] Xiao-Xin, Liao. (1993). *Absolute Stability of Nonlinear Control Systems*. Kluwer, China.

[86] Yakubovich, V.A. (1998). A quadratic criterion for absolute stability. *Dokl. Math.* **58**, No.1, 169–172.

List of Main Symbols

Symbol	Meaning	Page
$\|\cdot\|$	Euclidean norm	5
I	identity operator	
$\|A\|$	operator norm of A	5
$\alpha(A) \equiv \max_k Re\ \lambda_k(A)$		8
$g(A)$		10
$\lambda_k(A)$	eigenvalue of A	
$N(A)$	Frobenius norm of A	5
$\beta(A) \equiv \min_k Re\ \lambda_k(A)$		
$R_\lambda(A)$	resolvent of A	12
$\gamma_{n,p}$		11
A^*	operator conjugate to A	
$Tr\ A = Trace\ A$	trace of A	
$\sigma(A)$	spectrum of A	
$\rho(A, \lambda)$	distance between $\sigma(A)$ and λ	
$det(A)$	determinant of A	
$\mathbf{C}^n$	complex Euclidean space	
$\mathbf{R}^n$	real Euclidean space	
$(.,.)$	scalar product	
$\Omega(r)$		71
$R_+ = [0, \infty)$		
$\|\cdot\|_{L^p} = \|\cdot\|_{L^p(R_+, C^n)}$		26
$\|\cdot\|_C = \|\cdot\|_{C(R_+, C^n)}$		26
$L^p(R_+, \mathbf{C}^n)$		26
$C(R_+, \mathbf{C}^n)$		26

Index

Lecture Notes in Control and Information Sciences

Edited by M. Thoma and M. Morari

Further volumes of this series can be found on our homepage:
springeronline.com

Vol. 313: Li, Z.; Soh, Y.; Wen, C.
Switched and Impulsive Systems
277 p. 2005 [3-540-23952-9]

Vol. 312: Henrion, D.; Garulli, A. (Eds.)
Positive Polynomials in Control
313 p. 2005 [3-540-23948-0]

Vol. 311: Lamnabhi-Lagarrigue, F.; Loría, A.; Panteley, V. (Eds.)
Advanced Topics in Control Systems Theory
294 p. 2005 [1-85233-923-3]

Vol. 310: Janczak, A.
Identification of Nonlinear Systems Using Neural Networks and Polynomial Models
197 p. 2005 [3-540-23185-4]

Vol. 309: Kumar, V.; Leonard, N.; Morse, A.S. (Eds.)
Cooperative Control
301 p. 2005 [3-540-22861-6]

Vol. 308: Tarbouriech, S.; Abdallah, C.T.; Chiasson, J. (Eds.)
Advances in Communication Control Networks
358 p. 2005 [3-540-22819-5]

Vol. 307: Kwon, S.J.; Chung, W.K.
Perturbation Compensator based Robust Tracking Control and State Estimation of Mechanical Systems
158 p. 2004 [3-540-22077-1]

Vol. 306: Bien, Z.Z.; Stefanov, D. (Eds.)
Advances in Rehabilitation
472 p. 2004 [3-540-21986-2]

Vol. 305: Nebylov, A.
Ensuring Control Accuracy
256 p. 2004 [3-540-21876-9]

Vol. 304: Margaris, N.I.
Theory of the Non-linear Analog Phase Locked Loop
303 p. 2004 [3-540-21339-2]

Vol. 303: Mahmoud, M.S.
Resilient Control of Uncertain Dynamical Systems
278 p. 2004 [3-540-21351-1]

Vol. 302: Filatov, N.M.; Unbehauen, H.
Adaptive Dual Control: Theory and Applications
237 p. 2004 [3-540-21373-2]

Vol. 301: de Queiroz, M.; Malisoff, M.; Wolenski, P. (Eds.)
Optimal Control, Stabilization and Nonsmooth Analysis
373 p. 2004 [3-540-21330-9]

Vol. 300: Nakamura, M.; Goto, S.; Kyura, N.; Zhang, T.
Mechatronic Servo System Control
Problems in Industries and their Theoretical Solutions
212 p. 2004 [3-540-21096-2]

Vol. 299: Tarn, T.-J.; Chen, S.-B.; Zhou, C. (Eds.)
Robotic Welding, Intelligence and Automation
214 p. 2004 [3-540-20804-6]

Vol. 298: Choi, Y.; Chung, W.K.
PID Trajectory Tracking Control for Mechanical Systems
127 p. 2004 [3-540-20567-5]

Vol. 297: Damm, T.
Rational Matrix Equations in Stochastic Control
219 p. 2004 [3-540-20516-0]

Vol. 296: Matsuo, T.; Hasegawa, Y.
Realization Theory of Discrete-Time Dynamical Systems
235 p. 2003 [3-540-40675-1]

Vol. 295: Kang, W.; Xiao, M.; Borges, C. (Eds)
New Trends in Nonlinear Dynamics and Control, and their Applications
365 p. 2003 [3-540-10474-0]

Vol. 294: Benvenuti, L.; De Santis, A.; Farina, L. (Eds)
Positive Systems: Theory and Applications (POSTA 2003)
414 p. 2003 [3-540-40342-6]

Vol. 293: Chen, G. and Hill, D.J.
Bifurcation Control
320 p. 2003 [3-540-40341-8]

Vol. 292: Chen, G. and Yu, X.
Chaos Control
380 p. 2003 [3-540-40405-8]

Vol. 291: Xu, J.-X. and Tan, Y.
Linear and Nonlinear Iterative Learning Control
189 p. 2003 [3-540-40173-3]

Vol. 290: Borrelli, F.
Constrained Optimal Control of Linear and Hybrid Systems
237 p. 2003 [3-540-00257-X]

Vol. 289: Giarré, L. and Bamieh, B.
Multidisciplinary Research in Control
237 p. 2003 [3-540-00917-5]

Vol. 288: Taware, A. and Tao, G.
Control of Sandwich Nonlinear Systems
393 p. 2003 [3-540-44115-8]

Vol. 287: Mahmoud, M.M.; Jiang, J.; Zhang, Y.
Active Fault Tolerant Control Systems
239 p. 2003 [3-540-00318-5]

Vol. 286: Rantzer, A. and Byrnes C.I. (Eds)
Directions in Mathematical Systems
Theory and Optimization
399 p. 2003 [3-540-00065-8]

Vol. 285: Wang, Q.-G.
Decoupling Control
373 p. 2003 [3-540-44128-X]

Vol. 284: Johansson, M.
Piecewise Linear Control Systems
216 p. 2003 [3-540-44124-7]

Vol. 283: Fielding, Ch. et al. (Eds)
Advanced Techniques for Clearance of
Flight Control Laws
480 p. 2003 [3-540-44054-2]

Vol. 282: Schröder, J.
Modelling, State Observation and
Diagnosis of Quantised Systems
368 p. 2003 [3-540-44075-5]

Vol. 281: Zinober A.; Owens D. (Eds)
Nonlinear and Adaptive Control
416 p. 2002 [3-540-43240-X]

Vol. 280: Pasik-Duncan, B. (Ed)
Stochastic Theory and Control
564 p. 2002 [3-540-43777-0]

Vol. 279: Engell, S.; Frehse, G.; Schnieder, E. (Eds)
Modelling, Analysis, and Design of Hybrid Systems
516 p. 2002 [3-540-43812-2]

Vol. 278: Chunling D. and Lihua X. (Eds)
H_∞ Control and Filtering of
Two-dimensional Systems
161 p. 2002 [3-540-43329-5]

Vol. 277: Sasane, A.
Hankel Norm Approximation
for Infinite-Dimensional Systems
150 p. 2002 [3-540-43327-9]

Vol. 276: Bubnicki, Z.
Uncertain Logics, Variables and Systems
142 p. 2002 [3-540-43235-3]

Vol. 275: Ishii, H.; Francis, B.A.
Limited Data Rate in Control Systems with Networks
171 p. 2002 [3-540-43237-X]

Vol. 274: Yu, X.; Xu, J.-X. (Eds)
Variable Structure Systems:
Towards the 21st Century
420 p. 2002 [3-540-42965-4]

Vol. 273: Colonius, F.; Grüne, L. (Eds)
Dynamics, Bifurcations, and Control
312 p. 2002 [3-540-42560-9]

Vol. 272: Yang, T.
Impulsive Control Theory
363 p. 2001 [3-540-42296-X]

Vol. 271: Rus, D.; Singh, S.
Experimental Robotics VII
585 p. 2001 [3-540-42104-1]

Vol. 270: Nicosia, S. et al.
RAMSETE
294 p. 2001 [3-540-42090-8]

Vol. 269: Niculescu, S.-I.
Delay Effects on Stability
400 p. 2001 [1-85233-291-316]

Vol. 268: Moheimani, S.O.R. (Ed)
Perspectives in Robust Control
390 p. 2001 [1-85233-452-5]

Vol. 267: Bacciotti, A.; Rosier, L.
Liapunov Functions and Stability in Control Theory
224 p. 2001 [1-85233-419-3]

Vol. 266: Stramigioli, S.
Modeling and IPC Control of Interactive Mechanical
Systems – A Coordinate-free Approach
296 p. 2001 [1-85233-395-2]

Vol. 265: Ichikawa, A.; Katayama, H.
Linear Time Varying Systems and Sampled-data Systems
376 p. 2001 [1-85233-439-8]

Vol. 264: Baños, A.; Lamnabhi-Lagarrigue, F.;
Montoya, F.J
Advances in the Control of Nonlinear Systems
344 p. 2001 [1-85233-378-2]

Vol. 263: Galkowski, K.
State-space Realization of Linear 2-D Systems with
Extensions to the General nD ($n>2$) Case
248 p. 2001 [1-85233-410-X]

Vol. 262: Dixon, W.; Dawson, D.M.; Zergeroglu, E.;
Behal, A.
Nonlinear Control of Wheeled Mobile Robots
216 p. 2001 [1-85233-414-2]

Vol. 261: Talebi, H.A.; Patel, R.V.; Khorasani, K.
Control of Flexible-link Manipulators
Using Neural Networks
168 p. 2001 [1-85233-409-6]

Vol. 260: Kugi, A.
Non-linear Control Based on Physical Models
192 p. 2001 [1-85233-329-4]

Vol. 259: Isidori, A.; Lamnabhi-Lagarrigue, F.;
Respondek, W. (Eds)
Nonlinear Control in the Year 2000 Volume 2
640 p. 2001 [1-85233-364-2]

Vol. 258: Isidori, A.; Lamnabhi-Lagarrigue, F.;
Respondek, W. (Eds)
Nonlinear Control in the Year 2000 Volume 1
616 p. 2001 [1-85233-363-4]

Vol. 257: Moallem, M.; Patel, R.V.; Khorasani, K.
Flexible-link Robot Manipulators
176 p. 2001 [1-85233-333-2]

Printing: Krips bv, Meppel
Binding: Litges & Dopf, Heppenheim